CRC Series in Mathematical Models in Microbiology

Editor-in-Chief

Michael J. Bazin, Ph.D.

Microbial Population Dynamics

Editor

Michael J. Bazin, Ph.D.

Soil Microbiology: A Model of Decomposition and Nutrient Cycling

Author

O. L. Smith, Ph.D.

Physiological Models in Microbiology

Editors

Michael J. Bazin, Ph.D.
James I. Prosser, Ph.D.

Physiological Models in Microbiology

Volume I

Editors

Michael J. Bazin, Ph.D.
Senior Lecturer
Department of Microbiology
King's College
London, England

James I. Prosser, Ph.D.
Lecturer
Department of Genetics and Microbiology
University of Aberdeen
Aberdeen, Scotland

CRC Series in Mathematical Models in Microbiology

Editor-in-Chief
Michael J. Bazin, Ph.D.

CRC Press, Inc.
Boca Raton, Florida

Library of Congress Cataloging-in-Publication Data

Physiological models in microbiology

  Bibliography: p.
  Includes index.
  1. Micro-organisms—Physiology—Mathematical
models. 2. Biological models. I. Bazin, Michael J.
II. Prosser, James Ivor.
QR84.P46 1988   576'.11'0724   87-18330
ISBN 0-8493-5953-8 (set)
ISBN 0-8493-5954-6 (v. 1)
ISBN 0-8493-5955-4 (v. 2)

Direct all inquiries to CRC Press, Inc., 2000 Corporate Blvd., N.W., Boca Raton, Florida, 33431.

© 1988 by CRC Press, Inc.

International Standard Book Number 0-8493-5953-8 (set)
International Standard Book Number 0-8493-5954-6 (v. 1)
International Standard Book Number 0-8493-5955-4 (v. 2)

Library of Congress Card Number 87-18330
Printed in the United States

MATHEMATICAL MODELS IN MICROBIOLOGY

M. J. Bazin, Editor-in-Chief

This multivolume series will contain a selection of authoritative articles on the application of mathematical models to microbiology. Each volume will be devoted to a specialized area of microbiology and topics will be presented in sufficient detail to be of practical value to working scientists. A sincere attempt will be made to make the material useful for microbiologists with only moderate mathematical training. Under each title a variety of modeling techniques will be included, and both purely scientific and applied aspects of the subject will be covered.

The objectives of the series will be to introduce microbiologists familiar with the modeling approach to new models, methods of model construction, and analytical techniques, and to encourage those with limited mathematical backgrounds to incorporate modeling as an integral part of their research programs.

PREFACE

Volume I

The multi-volume series on **Mathematical Models in Microbiology** presents the application of mathematical models to the study of microbial processes and interactions. *Physiological Models in Microbiology* consists of two volumes. Volume I presents models of basic growth processes and the effects of environmental factors while models of secondary processes, spore germination, chemotaxis, surface growth, and microbial death are covered in Volume II.

Chapters 1 to 3 consider aspects of growth common to all microorganisms. The first presents a quantitative approach to the control and regulation of flux through metabolic pathways in which properties of the system may be derived from those of individual components. Chapter 2 describes new approaches to modeling microbial growth, particularly the application of thermodynamics, while Chapter 3 quantifies aspects of chemiosmosis and nutrient transport.

In Chapters 4 to 6, the effects of environmental factors on physiological processes are described. Chapter 4 discusses the effect of temperature on microbial growth and use of models in predicting shelf-life of food. Growth of photosynthetic organisms is considered in Chapter 5 and in the final chapter the effect of pH on microbial growth is described, with particular reference to wastewater treatment.

The articles illustrate both the predictive value of models and their potential in increasing our understanding of microbial processes and will hopefully be of interest to microbiologists in both pure and applied research in which a quantitative approach is required.

THE EDITORS

Michael J. Bazin, Ph.D., is a Senior Lecturer in the Department of Microbiology, King's College, University of London.

Dr. Bazin trained as a teacher at St. Luke's College, Exeter, and taught in secondary schools in England and the United States. He received his Ph.D. in Zoology from the University of Minnesota in 1968 after which time he pursued a postdoctoral traineeship in biomathematics at the University of Michigan.

Dr. Bazin has had wide research interests ranging from sexuality in blue-green algae to ethnic differences in skinfold thickness. His current major interests revolve around the application of mathematics to problems in biology and are directed chiefly towards theoretical biology and biotechnology.

James I. Prosser, Ph.D., is a Lecturer in the Department of Genetics and Microbiology, University of Aberdeen, Aberdeen, Scotland. Dr. Prosser received his B.Sc. degree in Microbiology from Queen Elizabeth College, University of London in 1972, and his Ph.D. from the University of Liverpool, Department of Botany in 1975. After three further years in the latter Department, as a Natural Environment Research Council Postdoctoral Fellow and Senior Demonstrator, he took up his current post at the University of Aberdeen.

Dr. Prosser's research interests fall into two main areas. The first is an investigation into the environmental factors affecting growth, activity, attachment, and inhibition of soil nitrifying bacteria. The second is the relationship between growth, branching, and secondary metabolite production by filamentous fungi and actinomycetes. The link between these two areas is the application of a quantitative approach and the use of theoretical models in the study of microbiological processes.

CONTRIBUTORS

Volume I

Arthur C. Anthonisen, Ph.D.
Consulting Engineer
MONTECO
Montgomery, New York

H. Kacser, Ph.D.
Department of Genetics
University of Edinburgh
Edinburgh, Scotland

Prasad S. Kodukula, Ph.D.
Union Carbide Corporation
Technical Center
South Charleston, West Virginia

Lee Yuan Kun, Ph.D.
Department of Microbiology
National University of Singapore
Singapore, Republic of Singapore

Y.-K. Lee, Ph.D.
Department of Microbiology
National University of Singapore
Singapore, Republic of Singapore

T. A. McMeekin, Ph.D.
Reader in Microbiology
Department of Agricultural Science
University of Tasmania
Hobart, Tasmania, Australia

Marcel Mulder, Ph.D.
Department of Biochemistry
B.C.P. Jansen Institute
University of Amsterdam
Amsterdam, The Netherlands

June Olley, D.Sc., Ph.D.
Senior Principal Research Scientist
CSIRO
Tasmanian Food Research Unit
Hobart, Tasmania, Australia

T. B. S. Prakasam, Ph.D.
Research and Development Laboratories
Metropolitan Sanitary District
Cicero, Illinois

D. A. Ratkowsky, Ph.D.
Principal Research Scientist
Division of Mathematics and Statistics
CSIRO
Battery Point, Tasmania, Australia

Dale Sanders, Ph.D.
Department of Biology
University of York
York, England

Teixeira de Mattos, Ph.D.
Lecturer
Laboratory of Microbiology
University of Amsterdam
Amsterdam, The Netherlands

Karel van Dam, Ph.D.
Professor
Department of Biochemistry
B.C.P. Jansen Institute
University of Amsterdam
Amsterdam, The Netherlands

Hans V. Westerhoff, Ph.D.
Visiting Scientist
Molecular Biology Laboratory
National Institutes of Health
Bethesda, Maryland

CONTRIBUTORS

Volume II

James D. Bryers, Ph.D.
Professor
Center for Biochemical Engineering
Duke University
Durham, North Carolina

Antonio Casolari
Libero Docente
Department of Microbiology
Stazione Sperimentale
Parma, Italy

Raymond Leblanc, Ph.D.
Professor
Department of Mathematics and
 Computer Sciences
Université du Québec à Trois-Rivières
Trois-Rivières, Québec, Canada

Gerald M. Lefebvre, Ph.D.
Professor
Department of Physics
Université du Québec à Trois-Rivières
Trois-Rivières, Québec, Canada

Gerald Rosen, Ph.D.
M. R. Wehr Professor
Department of Physics and Atmospheric
 Science
Drexel University
Philadelphia, Pennsylvania

Paul R. Rutter, Ph.D.
Senior Chemist
Minerals Processing Branch
British Petroleum Research Center
Middlesex, England

Brian Vincent, D.Sc., Ph.D.
Reader in Physical Chemistry
Department of Physical Chemistry
University of Bristol
Bristol, England

TABLE OF CONTENTS

Volume I

Volume II

Chapter 1

REGULATION AND CONTROL OF METABOLIC PATHWAYS

H. Kacser

TABLE OF CONTENTS

I. INTRODUCTION

In addressing the problems of control of metabolism in microorganisms it is useful to distinguish universal aspects, those that are common to all living systems, and specific aspects, those that are unique, or at least overwhelmingly represented, in cellular types. The universal aspects are represented by pathways and their enzymes leading to the production of energy, proteins, nucleic acids, "waste" products, and so on. There are, of course, differences between species, not only in the absence or presence of particular enzymes, but also in whole pathways and even complete areas of metabolic activity. We shall not discuss these differences, which are the proper concern of comparative biochemistry and evolution. The broad organization of any cell is sufficiently similar to any other that we can talk about *the* cell and use the information from studies in the field of biochemistry.

There are, however, some properties of microorganisms which largely set them apart. These are connected with the fact that microorganisms spend most or a most important part of their "life" in reproducing asexually or growing. As we shall see, this does affect their control properties. There are other differences, such as the absence of tissues and organs, which in some sense makes the analysis simpler, but in other ways has resulted in the development of mechanisms which are more elaborate.

We shall start by considering the cell and what universal features we can discern. The first is that the individual chemical transformations, which constitute the units of activity, are organized. This organization is not so much a spatial one, although this does exist, but is kinetic in nature. It results from the empirical fact that enzymes are coupled to one another by the metabolites they share. The product of one enzyme reaction is the substrate for another. This, in turn, donates its product to a third and so on. Pathways exist even in a homogeneous, well-stirred solution. The fact that some pathways form loops and cycles that diverge and converge, does not affect the conclusion that the rate through any chosen enzyme is affected by its enzymic neighbors insofar as they supply the substrate and remove the product. These neighbors have their own neighbors, etc., until you get to the "edges" of the system. For physiological conditions, then, we must consider the whole system. This kinetic organization of metabolism leads to the inevitable view that in principle, all the enzymes in the cell affect the rate through any one of them. Although there are broad-specificity enzymes, we shall mainly use the "one rate-one enzyme" fiction which is displayed in the metabolic charts so generously provided by the drug companies. Such metabolic maps, however, are kinetically barren. They merely represent the molecular skeleton. To flesh out these bones we must have quantitative kinetic data on every rate. Even if we had this unlikely mass of information, each rate being represented by often complex, algebraic expressions, our reconstruction would fail. To solve one equation we would have to solve all. The set of thousands of nonlinear simultaneous equations cannot be solved algebraically and are impractical to approach computationally.

Are then the labors of enzymologists in vain? To answer this we must decide what questions we want to ask. Questions about how metabolism is *controlled in the living cell* cannot be based on preconceived notions nor can one ignore the methodological impasse discussed earlier. Nevertheless, we find the literature and text books replete with such dubious approaches. A familiar question is "Which is the rate-determining enzyme in the pathway?" Another is "Which is the regulatory site in the pathway?" We hope to prove that the first question is an improper one and that the answer to the second may be uninformative. I shall replace these questions by others which take account of the systemic nature of the problem and which will allow us to answer them experimentally. The theoretical foundations of this approach and some of the experimental applications will be found elsewhere.[1-10]

II. THE CONTROL COEFFICIENTS

For any steady state, whether it is nongrowing and stationary or growing and exponential, the concentrations of enzymes and metabolites have time-invariant values. It is true that for growing and dividing unicellular organisms, this is only correct for a mean value taken over a number of cycles; however, unless we are interested in the details of the cell cycle, such mean values will be characteristic of the culture. We shall begin by considering nongrowing systems, such as liver cells or yeast cells in the later stages of alcohol fermentation where the rate of metabolism is substantially greater than the net synthesis of cell components.

For any simple section of a pathway,

$$\rightarrow S_i \overset{E_j}{\underset{v_j}{\rightarrow}} S_j \overset{E_k}{\underset{v_k}{\rightarrow}} S_k \overset{E_l}{\underset{v_l}{\rightarrow}} S_l \rightarrow \rightarrow P \qquad \text{(Scheme 1)}$$

the net rate, v, of any step is equal to that of all others and is equal to the system flux, J, through that part of the metabolism. This is achieved by the adjustments of the variables, i.e., the metabolite concentrations, until the conditions

$$\frac{dS_i}{dt} = 0 \qquad (1)$$

and

$$J = \frac{dP}{dt} = \text{Constant}$$

are satisfied. This will be irrespective of the concentrations and kinetic constants of the individual enzymes. At steady state all enzymes carry the same flux, J, and in that sense, all are equal. Similarly, if any one of the enzyme activities is reduced to zero (by virtue of the absence of the protein or the elimination of its catalytic activity), the flux will be zero. (We neglect the very small "spontaneous" rate of transformation.) Clearly, the flux is a function of all the enzymes. For these two conditions we cannot distinguish one enzyme from another. It does not, however, follow that variation in one of the enzymes has the same effect on the flux as that same variation in another enzyme.

Let us consider a small variation in the concentration of one of the enzymes, say $\delta[E_j]$. The result will be a change in the metabolite concentrations, first those flanking the enzyme, E_j, but eventually spreading throughout the whole system so that a new steady state is achieved with changes $\delta[S_i]$, $\delta[S_j]$ . . . and δJ. Since we are not concerned with the units in which these are expressed we shall consider the normalized or fractional changes

$$\frac{\delta E_j}{E_j}, \frac{\delta S_i}{S_i}, \frac{\delta S_j}{S_j}, \ldots \ldots \frac{\delta J}{J}$$

We can now ask the first quantitative question about control. How effective is the change of the enzyme in changing the variables? The ratio of effect/cause is a measure of this.

$$\frac{\delta J}{J} \bigg/ \frac{\delta E_j}{E_j} \qquad \text{measures the response of the flux}$$

and

$$\frac{\delta S_i}{S_i} \Big/ \frac{\delta E_j}{E_j} \qquad \text{measures the response of the metabolite}$$

These ratios, however, will not be independent of the actual change $\delta E_j/E_j$ since we are dealing with nonlinear relationships. In the limit, as $\delta E_j \to 0$, we obtain a differential expression independent of the value of δE_j, i.e., a coefficient.

A. Definitions

$$\frac{\partial J}{J} \Big/ \frac{\partial E_j}{E_j} = C_{E_j}^J$$

and

$$\frac{\partial S}{S} \Big/ \frac{\partial E_j}{E_j} = C_{E_j}^S$$

These are designated the flux control coefficient and the concentration control coefficient, respectively.[11] Since both S_i and J are functions of all the enzymes, the formulation represents a partial derivative, i.e., the change in the variable with respect to one of the parameters when all others are constant. If closed expressions for J and S_i were obtainable, partial differentiation of this would give an explicit formulation for the two types of coefficient. Only for special cases can such formulations be obtained.[1,5] Although the coefficients have been defined with respect to concentration changes, other parameters of the enzyme, such as the turnover number or the Michaelis constant, can be substituted. The coefficient defines the response to changes in activity.[11]

Two important points should be made about the control coefficients. They are dimensionless numbers whose magnitudes describe the effectiveness or importance of one particular enzyme in affecting the steady-state values of flux or metabolites. By comparing these magnitudes for the various enzymes in the pathway, a measure of their relative importance is obtained. Such quantitative comparisons should replace such notions as rate-limiting or nonlimiting enzymes. All enzymes will exert some control, the only question is how much. Since we have seen that the variables are functions of all the parameters of the system (enzymes, external substrates, etc.), the particular magnitude of such coefficients will depend on the values of these parameters. This means that alterations in other enzymes will affect the control which enzyme E_j exercises over the flux. Control coefficients are systemic properties and not simply a property of the enzyme whose role in the physiology they appear to describe.

It is a corollary of the systemic nature of the control coefficient that the same enzyme in the same pathway may have a different value in different situations. Thus, comparing two species, superficially alike, the control properties may not be the same. Perhaps we are inclined to place too much importance on the identity of the molecular anatomy (maps) and less on its kinetic organization which is less immediately accessible. Similarly, the same organism when placed in different environments will show different control behavior. A well-known example is the change from glycolysis to gluconeogenesis.

A second point arises from the definition of the control coefficients given above. Any particular flux or metabolite concentration will respond to changes in every enzyme of the system, i.e., there are as many control coefficients for the variable as there are enzymes. We should be clear that when we use the word enzyme this will include both soluble and membrane-bound proteins, translocators, and so on. Control coefficients may have positive or negative values. This is quite obvious for the concentration control coefficients. Any increase in, for instance, E_j will increase S_j, and subsequent metabolite pools, while an increase in E_k will decrease S_j, but increase the pools distal to E_k:

$$C_{E_j}^{S_j} = \text{positive}$$

$$C_{E_k}^{S_j} = \text{negative}$$

For flux control coefficients we must consider more elaborate pathways before negative coefficients arise.

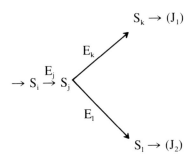

(Scheme 2)

There are now two fluxes, J_1 and J_2. Both will have positive coefficients with respect to E_j. On the other hand, E_k will affect J_1 positively but J_2 negatively, and vice versa for E_l (see Reference 6).

$$\mathbf{C}_{E_j}^{J_1} = \text{positive}; \quad \mathbf{C}_{E_k}^{J_1} = \text{positive}; \quad \mathbf{C}_{E_l}^{J_1} = \text{negative}$$

$$\mathbf{C}_{E_j}^{J_2} = \text{positive}; \quad \mathbf{C}_{E_k}^{J_2} = \text{negative}; \quad \mathbf{C}_{E_l}^{J_2} = \text{positive}$$

It should be noted that such negative flux control coefficients only arise if the two fluxes are truly independent (after the branch), i.e., if they exit the system or if the end products are irreversibly precipitated. Should the products of say J_1, even after many steps, reenter the pathway before the branching point at S_j, $C_{E_k}^{J_2}$ will be positive. Negative coefficients will therefore be less numerous than a cursory glance at the map may suggest.

In any actual biological system there are many fluxes, some stoichiometrically constrained and some related by competitive interactions for common substrates and, similarly, many intermediate metabolite pools. For any one of these variables the control coefficients give a quantitative measure of the extent to which particular enzymes can affect their steady-state values. Nothing has been said thus far about the magnitudes which these coefficients can and do take. It is indeed the aim of control analysis to determine these magnitudes because they describe in absolute and relative terms the distribution of control in the system.

It is a matter of experience (and of biochemical common sense) that changes in enzyme activity in a part of metabolism which is very distant from the pathway of interest will not affect the flux or any metabolite concentration. In our terminology this indicates that such control coefficients (whether positive or negative) will be exceedingly small. Taking all the enzymes in the cell, the vast majority will have negligible coefficients *for any one variable* (flux or metabolite), although they may be quite important (have high coefficients) for their own pathway. It is furthermore intuitively obvious that the major contribution to the control will be with respect to the enzymes directly involved in the pathway. It should, however, be noted that these may well include enzymes concerned with co-factor synthesis and other effector molecules, but we do not know where the pathway begins and ends. The practical problem then is to determine the control coefficients of this group of enzymes although we cannot *a priori* decide what comprises the members of that group.

B. The Summation Properties

The development of the systemic theory of control[1-4,6] has produced a number of theorems which represent some of the properties of enzyme systems. These theorems give information on the distribution of control as well as establish relationships which are useful in determining the values experimentally.

We have seen that any flux or metabolite concentration has as many control coefficients as there are enzymes in the system. This is a formidable number, for example, between 10^4 and 10^5.

It is easily shown[1,2,4] that if all the individual values of the flux control coefficients are added, the sum equals unity.

$$\sum_{i=1}^{n} C_{E_i}^{J} = 1 \quad \text{flux summation property} \tag{2}$$

This applies to any one particular flux and there are therefore as many flux summation properties (Equation 2) as there are fluxes in the system. This property may be thought of as meaning that for each flux there is unit amount of control in the system, which is distributed among all the enzymes. We have already argued that most of these n coefficients are negligibly small so that, for the remaining, m, numbers of the major control group

$$\sum_{i=1}^{m} C_{E_i}^{J} \approx 1 \tag{2a}$$

The number m is, of course, a matter of empirical evidence but is probably not much less than 10 or more than 100. The expected value for the average major coefficient will therefore lie between perhaps 0.1 and 0.01. This means that we would expect small changes in most enzymes to have little effect on the flux even for enzymes within the pathway. This seems to be borne out experimentally. It is, however, the distribution of control within the pathway which is of interest. We shall see some examples of this within the constraint of Equation 2a.

A similar summation relationship exists for metabolites.

$$\sum_{i=1}^{n} C_{E_i}^{S_j} = 0 \quad \text{concentration summation property} \tag{3}$$

We have referred to the positive and negative coefficients of metabolite S_j. Inspection of the section of the pathway (Scheme 1) shows that all the concentration control coefficients with respect to enzymes proximal to S_j will be positive and those distal to S_j will be negative. This refers, of course, to the case when the net flux is left to right. (Should there be a reversal of flow, as occasionally may happen, the signs of the coefficients will be reversed.) As before, the concentration of a particular pool will be quite insensitive to movements of distant enzymes and the canceling out of positive and negative coefficients (Equation 3) is mainly due to enzyme changes in its neighborhood. One consequence of this would be that equal, simultaneous changes in enzyme activities in a pathway (coordinate changes) would leave all the pools undisturbed.

C. Experimental Estimates

The most direct way to determine the control coefficients is implied in their definition.

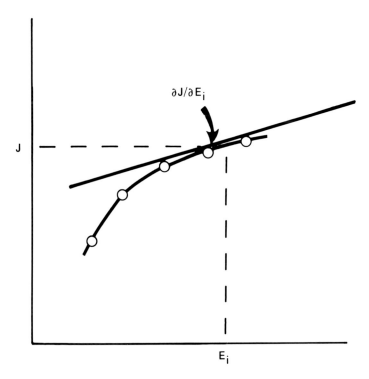

FIGURE 1.

$$C_{E_i}^J = \frac{\partial J}{J} \bigg/ \frac{\partial E_i}{E_i}$$

can be written as:

$$= \frac{\partial J}{\partial E_i} \cdot \frac{E_i}{J}$$

i.e., it is the slope of the J vs. E_i relationship multiplied by the scaling or normalizing factor E_i/J. Since infinitesimal changes cannot be brought about or measured, we have to determine a number of finite changes and interpolate between them (Figure 1). For any given value of E_i the slope can be estimated and, after scaling, will give the value of the coefficient at that value of E_i. The same methodology can be used for determining $C_{E_i}^{S_j}$ values. Some examples of this method and the means of achieving it are given in Figure 2.

An important aspect of the control coefficient is evident from the data in Figure 2. In all cases the slope of the relationships changes with the value of the enzyme activity and so does the value of the control coefficient. Since the summation property is true for any constellation of enzymes, this implies that a change in the value of the coefficient consequent upon a change in the activity of the enzyme will result in changes in all other coefficients, although no changes in other enzymes have taken place. If such changes (and particularly large changes) have taken place at some point (by induction or repression, mutation, effector interaction), a redistribution of the control in the pathway will be the result. It may well invert the relative importance of different enzymes.

For a simple pathway consisting of unsaturated enzymes, an explicit expression for the flux control coefficient can be obtained.[1,5]

$$C_{E_i}^J = \frac{1/e_i}{1/e_1 + 1/e_2 + \dots 1/e_i + \dots 1/e_n} \tag{4}$$

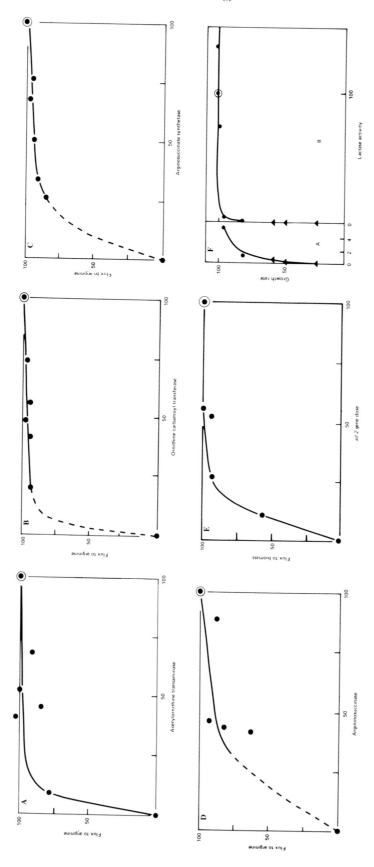

FIGURE 2.   Flux-enzyme relationships. In all relationships the circled points represent wildtype, scaled to 100 for both enzyme activity and flux. (A to D) *Neurospora* heterokaryons. Each figure represents a series of heterokaryons with different nuclear ratios of null to wildtype nuclei. They were constructed by varying the input ratios of the respective mutant and wildtype conidia.[9]   (E) *saccharomyces* mutant at the *ad-2* locus. Triploid and tetraploid strains were constructed having various doses of null and wildtype alleles of the *ad-2* locus, while the rest of the genome was isogenic. The locus specifies phosphoribosyl aminoimidazole carboxylase. The flux to biomass is the exponential growth constant.[26]   (F) *E. coli* β-galactosidase mutants. Variants of lactase activity and their respective growth rates. ●, *lacz* mutants. ▲, *ebg* mutants. (A) Expanded scale, 0—6, of (B).[27]   Other examples in higher eukaryotic systems can be found in References 5, 17, 28, and 29.

where e_i is an expression for the enzyme activity of the form

$$e_i = [E_i] \, k_{cat} \cdot K_{eq}/K_m \qquad (5)$$

Although representing a very simplified system, it shows all the aspects discussed so far, namely the summation property, the change of magnitude of the coefficient with changes in e_i, and its change with changes in any of the other enzyme activities. All these aspects have been confirmed experimentally investigating oxidative phosphorylation in mitochondria,[7,8] tryptophan catabolism in rat liver,[17] and part of glycolysis in an in vitro system.[30]

Direct determination of control coefficients, as exemplified in the cases of Figure 2, depends on the ability to modulate enzyme activity by the use of mutants, heterokaryons, or induction/repression of particular enzymes. This is not always possible. We shall now discuss other methods which extend the use of control analysis.

Genetic manipulation technology is the most recent method to be used to alter specific enzyme concentrations in suitable hosts. In microorganisms such as *Escherichia coli* or yeast, a freely replicating vector or plasmid carrying the gene for a particular enzyme will result in increased enzyme activity at that step. If different numbers of vectors can be maintained by some alterations of the vector properties or if a variable expression vector is used, a series of concentration-flux data is obtainable. Unlike the heterokaryon method used in *Neurospora*, or polyploid yeast strains which investigate decreases in enzyme concentrations, plasmid technology can explore the effects of increased enzyme activities. This has been used by Niederberger and his colleagues[32] investigating tryptophan synthesis in *Saccharomyces cerevisiae*. As was shown previously,[23] reducing individual enzymes in tetraploids showed low coefficients for each of the five principal enzymes of the pathway. Similarly and predictably, increasing individual enzyme concentrations by means of plasmids was equally ineffective in changing the flux significantly. Construction of a vector with all the genes of the five enzymes, however, showed a nearly proportional response in the tryptophan synthesis (summation property).

Another application[31] uses vectors containing the gene for citrate synthase attached to the *tac* promoter in an *E. coli* host lacking synthase activity. By growing the bacteria with different concentrations of IPTG, a series of different enzyme concentrations, spanning both sides of the wild-type levels could be achieved. Measurement of the flux in the Krebs cycle and glyoxalate shunt showed considerable differences in control coefficients when comparisons of glucose-acetate medium and acetate alone were made. The results emphasize the systemic nature of these coefficients.

III. THE ELASTICITY COEFFICIENTS

For these we must turn to the properties of individual enzymes. Unlike with classical enzymology, we shall not be concerned with the detailed kinetic scheme of each enzyme, the mechanisms, transition states, and so on. For our purpose we shall only consider those aspects of their kinetics which are relevant to the control behavior of the system. We shall pursue the questions of what these aspects are and how they affect the control coefficients. It must be true that any systemic property will be due to the properties of its constituent parts and the interactions between them.

We have seen that changes which occur when the activity of a particular enzyme is modulated are transmitted via the movement of the metabolites which are the links between the different enzymes. When the new steady state has been reached, a new constellation of intermediate pools will be present with concentrations which satisfy the condition that a different flux is sustained. This means that the rate through every enzyme is altered and, in the case of a simple pathway, altered by the same amount.

$$\delta v_i = \delta v_j = \delta v_k = \text{-----} = \delta J \tag{6}$$

The metabolite concentrations, on the other hand, will not have moved by the same amount. Take two adjacent enzymes in a pathway, E_k and E_l in Scheme 1; the first is relatively unsaturated, the second highly saturated. Let there be a finite change in E_j resulting in change δJ (Equation 6). It is obvious that the change in concentration of the substrate to the second enzyme (S_k) must be much greater than that of the substrate to the first enzyme (S_j) in order that both rates show the same change. In general:

$$\delta S_i \neq \delta S_j \neq \delta S_k \text{ ------} \tag{7}$$

We shall therefore consider the changes around each enzyme. Let us take the simple scheme (Scheme 1) where each enzyme is flanked by only one substrate and one product. The arguments which follow, however, apply to any reaction with several reactants as well as effectors from other pathways. We again consider a change in the activity of enzyme E_j, but now focus our attention on the rate between S_k and S_l catalyzed by enzyme E_l. This enzyme "sees" only the molecules which interact directly with it and the rate will change only if the concentrations of these molecules change, irrespective of the initial cause of such a change. We are taking an enzyme's-eye-view of the events in the system. This is, of course, an enzymological approach, but we are interested only in the response of the catalyzed rate due to changes in the molecular partners in the reaction at their operating concentrations. We have isolated (at least in our minds) the enzyme from the rest of the system and surrounded it with the steady-state concentrations of its metabolites.

To predict what change in the rate will occur we must know how the rate would respond to each participating molecule separately. Again, we use normalized and infinitesimal changes. In our simple example the rate is a function of S_k, S_l, and E_l only, the particular algebraic form of this function being, for the moment, of no interest.

$$v_l = f(S_k, S_l, E_l)$$

The effect of changes in S_k is then defined by

$$\left(\frac{\partial v_l}{v_l} \bigg/ \frac{\partial S_k}{S_k} \right)_{S_l \, E_l} = \epsilon_{S_k}^{v_l}, \quad \text{or simply} = \epsilon_k^l$$

Similarly

$$\left(\frac{\partial v_l}{v_l} \bigg/ \frac{\partial S_l}{S_l} \right)_{S_k, \, E_l} = \epsilon_l^l$$

and

$$\left(\frac{\partial v_l}{v_l} \bigg/ \frac{\partial E_l}{E_l} \right)_{S_l, \, S_k} = \epsilon_{E_l}^l$$

We therefore have three responses of the rate to each of the molecular species which make up the reaction scheme. It should be noted that for each response the other concentrations are held constant.

These are designated the elasticity coefficients and refer to the isolated rate in the steady-state milieu. They are again partial derivatives and if the exact form of the rate expression

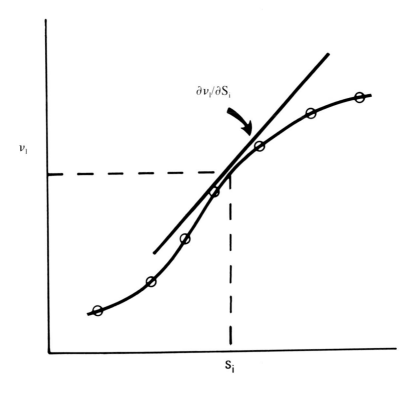

$\partial v_l / \partial S_i$

v_l

S_i

FIGURE 3.

were known and the absolute concentrations determinable, partial differentiation would lead to the value of the elasticity coefficients, As far as the third elasticity coefficient is concerned, its value will usually be equal to 1, since for most reactions the rate is proportional to enzyme concentration. (The exceptions to this would be if there are significant protein-protein interactions such as might occur in monomer $\leftrightarrows$ oligomer equilibria if their activities differed or if there are other heterologous enzyme-enzyme interactions.)

The first elasticity coefficient, ϵ_k^l, will be recognized as the slope of rate vs. its substrate concentration, as in the familiar Michaelis constant determinations except that it is the slope at one particular steady-state value of S_k and in the presence of all other steady-state metabolites (Figure 3). The magnitudes of the individual ϵ values will, apart from concentrations, depend on the kinetic constants of the enzyme. Here we need not know these since, in principle, elasticity coefficients can be determined empirically. In any case, for the moment we are simply concerned with defining them as functional responses.

If there are more reactants, products, and effectors, there would be additional elasticity coefficients. Each rate in the system, then, is represented by a number of elasticity coefficients which describe the potential response to changes in the concentrations. The values will be positive (with respect to substrates) or negative (with respect to products) and can take any value from 0 to ∞. This local response profile of any rate will determine how a change is transmitted. Since the individual rates represent the functional units of the kinetic system, the values of all the elasticity coefficients must jointly determine the net result of any change in the activity of any enzyme, which we have identified with the control coefficient.

It is worth stressing at this point that although the algebraic form of both types of coefficients, control and elasticity, are identical, i.e., the ratio of two infinitesimal changes representing the partial derivative of a function, they are conceptually and functionally quite different. The elasticity coefficient is a local property describing the response of the isolated rate, frozen in the steady state, to changes in each of the participating molecules. No interaction with the rest of the system is involved. It describes quantitatively how this step would respond in any system. The control coefficient, on the other hand, is a global property describing the net outcome of a change in the parameters of a single enzyme on a systemic variable. The effects of such a change reverberate through the whole system affecting every step, no longer isolated and responding to all movements of the metabolites and enzymes. The magnitude of the control coefficient is therefore not only dependent on the enzyme to which it refers but on all the others in the system. In what follows we shall show how the global properties are generated by the local ones.

Taking any rate in the system with a number of reactants, the change in this rate will be given by the simultaneous contributions of each of the reactants' changes. Provided the changes are small, it is simply the sum of all the contributions.

$$\frac{\delta v_i}{v_i} = \epsilon_1^i \frac{\delta S_1}{S_1} + \epsilon_2^i \frac{\delta S_2}{S_2} + \epsilon_3^i \frac{\delta S_3}{S_3} + \ldots \epsilon_{E_i}^i \frac{\delta E_i}{E_i} \qquad (8)$$

Each rate will be represented by such a sum of terms depending on the number of participating metabolites. There will be positive and negative ϵ values and the $\delta S_i/S_i$ terms may be negative or positive. In principle, we can write down a complete set of equations for every rate in the system. We will restrict ourselves to a very simple but complete system consisting of three enzymes only which will show the general relationship between the two types of coefficients.

$$X_0 \overset{E_1}{\rightarrow} S_1 \overset{E_2}{\rightarrow} S_2 \overset{E_3}{\rightarrow} X_3 \qquad \text{(Scheme 3)}$$

We assume constant concentrations for X_0 and X_3 which ensure that the system will evolve to a steady state irrespective of the initial conditions and the nature and concentrations of the enzymes. At steady state

$$dS_i/dt = 0$$

$$v_1 = v_2 = v_3 = J = dX_3/dt = \text{constant}$$

X_3 = constant (including zero) is not inconsistent with J = constant if some mechanism outside the system is assumed to operate holding X_0 and X_3 at constant concentrations. These are the same assumptions as those made for single enzyme assays. In practice they can be approximated sufficiently for the period of the experiment.

The elasticity coefficients in the system will be

$$\epsilon_0^1, \epsilon_1^1, \epsilon_1^2, \epsilon_2^2, \epsilon_2^3, \epsilon_3^3, \epsilon_{E_1}^1, \epsilon_{E_2}^2, \epsilon_{E_3}^3$$

Of these, $\epsilon_{E_i}^i = 1$ and ϵ_0^1 and ϵ_3^3 will not enter any equations of the type (Equation 8) since they will be multiplied by zero ($dX_0/X_0 = 0$, $dX_3/X_3 = 0$). We then have three equations for the changes in the rates consequent upon some unspecified change having occurred which will take the system from one steady state to a neighboring one.

$$\frac{\delta v_1}{v_1} = \epsilon_1^1 \frac{\delta S_1}{S_1} + \frac{\delta E_1}{E_1} \qquad (9)$$

$$\frac{\delta v_2}{v_2} = \epsilon_1^2 \frac{\delta S_1}{S_1} + \epsilon_2^2 \frac{\delta S_2}{S_2} + \frac{\delta E_2}{E_2} \qquad (10)$$

$$\frac{\delta v_3}{v_3} = \epsilon_2^3 \frac{\delta S_2}{S_2} + \frac{\delta E_3}{E_3} \tag{11}$$

At steady state the three changes must be equal to one another and equal to the change in the system flux $\delta J/J$. We can now return to the questions originally posed: What is the effect on the flux of changing one of the enzymes? (What is the value of the flux control coefficient?) and What is the distribution of control between the three enzymes? (What are the ratios of the control coefficient values?)

Let enzyme 1 be changed ($\delta E_1/E_1$) while leaving E_2 and E_3 at their original values ($\delta E_2/E_3 = 0$, $\delta E_3/E_3 = 0$). Dividing all three equations by $\delta E_1/E_1$ and taking the changes to the limit, $\delta E_1 \to 0$, we obtain for Equation 9

$$\frac{\partial J}{J} \Big/ \frac{\partial E_1}{E_1} = \epsilon_1^1 \frac{\partial S_1}{S_1} \Big/ \frac{\partial E_1}{E_1} + 1$$

which is seen to be

$$C_{E_1}^J = \epsilon_1^1 C_{E_1}^{S_1} + 1 \tag{9a}$$

and similarly

$$C_{E_1}^J = \epsilon_1^2 C_{E_1}^{S_1} + \epsilon_2^2 C_{E_1}^{S_2} \tag{10a}$$

$$C_{E_1}^J = \epsilon_2^3 C_{E_1}^{S_2} \tag{11a}$$

Eliminating the concentration control coefficients from the three equations yields

$$C_{E_1}^J = \frac{\epsilon_2^3 \epsilon_1^2}{\epsilon_2^3 \epsilon_1^2 - \epsilon_2^3 \epsilon_1^1 + \epsilon_2^2 \epsilon_1^1} \tag{12}$$

This demonstrates that the global flux control coefficient is generated by all the elasticity coefficients acting in the system. We can repeat this by considering changes in E_2 and E_3, respectively, obtaining the other two control coefficients.

$$C_{E_2}^J = \frac{-\epsilon_2^3 \epsilon_1^1}{\epsilon_2^3 \epsilon_1^2 - \epsilon_2^3 \epsilon_1^1 + \epsilon_2^2 \epsilon_1^1} \tag{13}$$

$$C_{E_3}^J = \frac{\epsilon_2^2 \epsilon_1^1}{\epsilon_1^3 \epsilon_1^2 - \epsilon_2^3 \epsilon_1^1 + \epsilon_2^2 \epsilon_1^1} \tag{14}$$

For larger and more complex systems the formulations would similarly contain all the elasticity coefficients as rather complex functions. A simple matrix method has been described for any system.[12] It should be noted that the summation property is clearly obeyed for the three coefficients.

For the calculation of flux control coefficients such large formulations are clearly not practical. They do, however, demonstrate a relationship between certain coefficients which is of considerable experimental use.

A. The Connectivity Properties

If we divide Equation 12 by Equation 13, we obtain:

$$C_{E_1}^J \Big/ C_{E_2}^J = -\epsilon_1^2 \Big/ \epsilon_1^1 \tag{15}$$

and Equation 13 by Equation 14:

$$C_{E_2}^J / C_{E_3}^J = -\epsilon_2^3 / \epsilon_2^2 \tag{16}$$

In Equation 15 the metabolite shared between enzymes E_1 and E_2 is S_1 with respect to which the two elasticity coefficients are defined. It is similar for S_2 in Equation 16. It is thus possible to obtain a measure of the relative values (importance) of two adjacent enzymes by determining only two elasticity coefficients no matter how complex or large the formulation for each may be. If successive pairs of elasticity coefficients are determined, the successive ratios of the control coefficients can be calculated along the whole pathway. An experimental method for in vivo determination of elasticity coefficients is given in Reference 4. If the absolute value of one of the latter is then determined, we obtain the absolute values of all. Equation 15 can also be written as:

$$C_{E_1}^J \, \epsilon_1^1 + C_{E_2}^J \cdot \epsilon_1^2 = 0 \tag{15a}$$

If more than two enzymes are interacting with one substrate (divided pathways, effector interactions, etc.), each interacting enzyme adds another term to the sum in Equation 15a. In general:[1,4,12]

$$\sum_{i=1}^{n} C_{E_i}^J \cdot \epsilon_S^i = 0 \quad \text{flux connectivity property} \tag{17}$$

The concentration control coefficients (obtainable from Equations 9a to 11a) have similar but different connectivity relations.[13] These are given by:

$$\sum_{i=1}^{n} C_{E_i}^{S_j} \cdot \epsilon_{S_k}^i = -\delta_{jk} \quad \text{concentration connectivity property} \tag{18}$$

where δ_{jk} is the Kronecker which equals 0 if $j \neq k$ and equals 1 if $j = k$. These relationships have been used, in part, to determine the control distribution in mitochondrial respiration.[8,14,15]

A third method to determine control coefficients makes use of inhibitors of particular enzymes applied to the system by the experimenter.[1] If such an external inhibitor (or external activator) is applied and if methods are available to measure the effect on the flux within the system, we can obtain a relationship which, superficially, looks like the inverse of the flux-enzyme relationship given in Figure 1 (see Figure 4).

From this, the slope can be estimated at any value of I and a coefficient calculated

$$\frac{\partial J}{J} \Big/ \frac{\partial I}{I} = R_I^J \tag{19}$$

We designate this the response coefficient.[1] We may be interested in the behavior of the system when some drug is applied. Of more fundamental importance is the use of such inhibitors to probe the enzyme as a means of determining the control coefficient in the absence of such an inhibitor. The effect is twofold. The first is on the activity of the enzyme and the second is the effect of the change in enzyme activity on the system.

The effect on the activity of the enzyme is measured by an elasticity coefficient.

$$\frac{\partial v_i}{v_i} \Big/ \frac{\partial I}{I} = {}^{\kappa}\epsilon_I^i \tag{20}$$

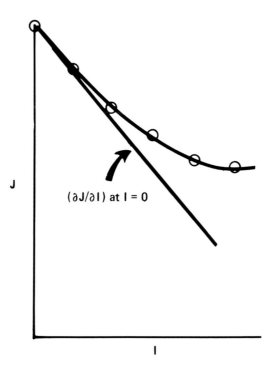

$(\partial J/\partial I)$ at $I = 0$

FIGURE 4.

This elasticity coefficient is distinguished by a superscript $^{\kappa}\epsilon$ to distinguish it from the internal elasticities which are with respect to variable pools of the metabolism. $^{\kappa}\epsilon$ does not enter the connectivity relationships of Equations 17 and 18.

The effect of the change in enzyme activity on the flux, consequent on the change in inhibitor, is the normal flux control coefficient.

$$\frac{\partial J}{J} \bigg/ \frac{\partial E_i}{E_i} = C^J_{E_i} \tag{21}$$

It is easily shown[1] that the overall response, the combined response coefficient, is the product of the two expressions above

$$R^J_I = C^J_{E_i} \cdot \epsilon^i_I \quad \text{combined response} \tag{22}$$

Hence:

$$C^J_{E_i} = R^J_I / \epsilon^i_I \tag{22a}$$

This can be written (using Equations 19, 20, and 21) as:

$$C^J_{E_i} = \frac{\partial J}{\partial I} \cdot \frac{I}{J} \bigg/ \frac{\partial v_i}{\partial I} \cdot \frac{I}{v_i}$$

Eliminating I, we obtain the ratios of two slopes (normalized only with respect to J and v_i)

$$= \frac{\partial J/J}{\partial I} \bigg/ \frac{\partial v_i/v_i}{\partial I}$$

Since the external inhibitor was only used to obtain the two responses, we are interested in the extrapolation to zero inhibitor.[7,8]

$$C_{E_i}^J = \left(\frac{\partial J/J}{\partial I}\right) \text{ s.s. at } I = 0 \bigg/ \left(\frac{\partial v_i/v_i}{\partial I}\right) S_i, S_j \text{ at } I = 0 \qquad (23)$$

The first term is obtained from the initial slope of the flux vs. I relationship (see Figure 4). The second term can be obtained by in vitro determination with the isolated enzyme but with all the steady-state milieu of the whole system. This may be difficult to achieve experimentally. Alternatively, it is obtainable by the partial derivative of the rate expression for v_i with respect to I, i.e., the kinetics of the inhibited reactions.

For example, the second term will be

1.   For irreversible inhibitors: $1/I_{max}$, where I_{max} equals the amount of inhibitor when all the enzyme is inhibited
2.   For noncompetitive inhibition: $-1/K_i$
3.   For competitive inhibition of a reversible reaction: $-1/K_i(1 + S_i/K_{si} + S_j/K_{sj})$

and more complex expressions for other types of inhibition.

This method has been successfully applied to the adenine nucleotide translocator in mitochondrial respiration[7,8] using carboxyatractyloside as the inhibitor and to tryptophan transport into hepatocytes using phenylalanine as the inhibitor.[17]

The combined response (Equation 22) discloses an important quantitative aspect. How much a given flux will be affected by an effector will depend not only on the kinetics of the enzyme ($^\kappa\epsilon$), but also on the systemic property of the whole system (C). It is often, quite wrongly, suggested that the identification of a control site (i.e., the step where an effector acts) is evidence that control is exercised there. It is evident that a high elasticity coefficient (i.e., a strong effect on the isolated or in vitro enzyme) may be ineffective by being multiplied by a low control coefficient. A regulated enzyme is no guarantee that in vivo it regulates the flux.

Since in most useful experiments the external nutrient concentrations are held constant by the experimenter, these substances are external effectors and play the same role as the inhibitors just discussed. Their experimental variation therefore provides a means to determine the control coefficients of the first steps (usually transport rates).

$$R_{X_0}^J = C_{E_1}^J \cdot {}^\kappa\epsilon_{X_0}^1 \qquad (24)$$

This has been used for the determination of the transport coefficient of tryptophan in hepatocytes.[17]

A fourth method uses computer simulation. While the simultaneous equations of any but the simplest system cannot be solved algebraically, it is quite possible to set up a reduced complex model and insert reasonable parameters obtained from measured fluxes and enzymic values. Provided the simulated model corresponds reasonably with the in vivo system (and this will not require a detailed knowledge of all the kinetics), a very simple set of instructions to modulate each enzyme in turn will print out the values of the control coefficients.[1,16,33] These experiments in numero, together with some experimentally determined coefficients, are likely to become increasingly useful as a means of investigating the more complex parts of metabolism where intuition is not likely to lead to a sound basis for experimental work.

IV. NEGATIVE FEEDBACK

We have so far discussed the various methods of determining the values of control coefficients for any chosen step, provided experimental modulation methods are available. We have also shown (Equations 12 to 14) how these values arise from the kinetics of the individual steps represented by the values of the elasticity coefficients. Longer and more complex pathways would generate very elaborate expressions. What these three expressions do not reveal readily are certain important effects which are generally found in metabolic systems, namely feedback inhibitions and repressions. In Scheme 4 we analyze such a situation:

$$\epsilon_2^1$$

$$r - \theta - - - - - \gamma$$

$$E_1 \qquad E_2 \qquad E_3$$

$$X_0 \rightarrow S_1 \rightarrow S_2 \rightarrow X_3 \qquad \text{(Scheme 4)}$$

In addition to elasticity coefficients previously described, the effect of S_2 on the rate v_1 must be represented by a new elasticity coefficient, ϵ_2^1, which is negative. When the algebraic equations are solved we obtain for the control coefficient of the first step:

$$C_{E1}^J = \frac{\epsilon_2^3 \epsilon_1^2}{\epsilon_2^3 \, \epsilon_1^2 - \epsilon_2^3 \, \epsilon_1^1 + \epsilon_2^2 \, \epsilon_1^1 - \epsilon_2^1 \, \epsilon_1^2} \qquad (25)$$

The additional term in the denominator is positive (as are all the others) and therefore the new C_{E1}^J value is smaller than the one without feedback. This means that while the loop is operating, the inhibited step has lost some of its importance. In the limit, when we have a very high feedback elasticity, i.e., when the fourth term is $>>$, then the sum of the first three, $C_{E1}^J \rightarrow 0$. All steps within the loop and proximal to it will have reduced control coefficients. On the other hand, steps distal to the signal will have increased coefficients. In our simple example:

$$C_{E3}^J = \frac{\epsilon_2^3 \epsilon_1^2 - \epsilon_2^1 \, \epsilon_1^2}{\epsilon_2^3 \, \epsilon_1^2 - \epsilon_2^3 \, \epsilon_1^1 + \epsilon_2^2 \, \epsilon_1^1 - \epsilon_2^1 \, \epsilon_1^2} \qquad (26)$$

In the limit (large ϵ_2^1), $C_{E3}^J \rightarrow 1$. The net effect of a feedback loop, then, is to transfer control to the step(s) distal to the signal. The summation property applies, of course, to this situation so that it is still true that

$$\sum_{i=1}^{n} C_{E1}^J = 1$$

but with a new distribution of values.

Once again, the systemic nature of control is demonstrated in this example. Although the control site is the enzyme E_1, all the control coefficients within the loop will have reduced values and those distal to the signal increased ones. The strength of the effect is seen to be dependent on the relative values of the feedback to the various elasticity terms which refer to all the other enzymes. In no sense can the inhibitable enzyme be said to be the only control in the pathway.

In the above example the effect of S_1 was described as an inhibition of the activity of enzyme (1). The same general relationship applies to the case where S_2 is a signal for the synthesis of the enzyme. Throughout the treatment, enzyme concentrations were taken as parameters, although in reality they are maintained by a balance of synthesis and degradation.

If, therefore, there is an effect of a pathway product on the synthesis of any enzyme, as is now widely demonstrated, the equations will have to be replaced by another set in which the enzyme(s) are variables, i.e., some function of the metabolites. These functions are not yet quantifiable (see, however, References 18 and 19) but the general case has been treated.[1]

The net behavior of such a repression/induction system is, however, the same as discussed above. A strong repression loop will transfer control to steps distal to the signal.

In many microorganisms some enzymes of a pathway are subject to coordinate repression, i.e., they change concentrations in, more or less, the same proportions. In some fungi there appears to be a mechanism which controls the concentrations of all amino acid synthetic enzymes in a unitary fashion. Reference to Equation 2a suggests that if all the enzymes of a pathway (which will usually be included in the major control group) are changed coordinately, the flux will respond by about the same factor as the enzyme changes. It may therefore be thought that this mechanism has evolved to control the flux. Experimental evidence[20] suggests that this is unlikely to be generally true. In *Neurospora*, a strong negative feedback inhibition of arginine on an early enzyme of the pathway virtually eliminates the effects of large changes in the enzyme concentrations of the whole pathway, elicited by the coordinate repression loop. The same situation appears to operate in the tryptophan pathway of yeast.[21-23] Evolutionary speculations should be moderated by experimental evidence.

V. GROWING SYSTEMS

The treatment in the preceding pages was concerned with nongrowing systems which can reach stationary steady states. Most mammalian biochemists will deal experimentally with these conditions. Microbial biochemists, on the other hand, often deal with exponentially growing systems (whether cellular or mycelial) which, in some respects, behave differently. In particular, it is no longer true that at the steady state the rates through each enzyme in a pathway are equal to one another.

$$v_1 \neq v_2 \neq v_3 \neq \ldots\ldots$$

To show this we can set up the equations for a growing system in the following manner. Let S be a metabolite within a system being produced by a flux J_1 and removed by J_2 where J_1 and J_2 are the instantaneous fluxes/unit volume,

$$\xrightarrow{J_1} S \xrightarrow{J_2}$$

and S is the concentration, i.e., the amount/unit volume. J_1 and J_2 will be represented by some function involving S as well as other metabolites and associated kinetic constants.

$$J_1 = E_1 \cdot f(S, t, K_m, \text{etc.})$$

Let the total volume of the system be V, then the total amount of S is

$$S_T = S \cdot V$$

and the total fluxes:

$$J_{1T} = J_1 \cdot V$$

$$J_{2T} = J_2 \cdot V$$

If the system grows (V increases), the instantaneous rate of change of S_T will be given by the differences between the flux in and the flux out. Starting from any initial conditions, all concentrations and hence fluxes may change as well as the volume.

$$\frac{dS_T}{dt} = J_1 \cdot V - J_2 \cdot V$$

$$\frac{d(S \cdot V)}{dt} = V(J_1 - J_2)$$

$$V\frac{dS}{dt} + S\frac{dV}{dt} = V(J_1 - J_2) \tag{27}$$

This gives

$$\frac{dS}{dt} = J_1 - J_2 - S\left(\frac{1}{V}\frac{dV}{dt}\right) \tag{28}$$

The term

$$\frac{1}{V}\frac{dV}{dt} = \frac{d\ln V}{dt}$$

is the slope of the lnV (or ln weight, for convenience) vs. time relation. It is, in fact, the growth constant of the system at any time, not necessarily constant.

A. The Steady State

The steady state is defined when the system grows exponentially ($d\ln V/dt = k$) and when all metabolite concentrations and all fluxes have stationary values ($dS/dt = 0$, J_1 and J_2 constant). Equation 28 then becomes:

$$0 = \bar{J}_1 - \bar{J}_2 - \bar{S} \cdot k \tag{29}$$

or

$$\bar{J}_1 = \bar{J}_2 + \bar{S} \cdot k \tag{29a}$$

The product $\bar{S} \cdot k$ (which has the dimensions of a flux) can be regarded as a flux of the metabolite into the expanding volume J^x.

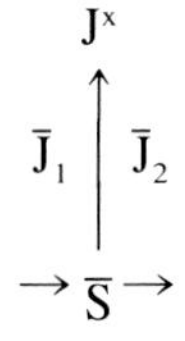

$$J^x$$
$$\uparrow$$
$$\bar{J}_1 \Big\vert \bar{J}_2$$
$$\rightarrow \bar{S} \rightarrow$$

(Scheme 5)

The input flux, at exponential steady state, is balanced by the sum of the output flux and the expansion flux. If J_2 is measured, by some means, J_1 can be calculated if the steady state concentration, $\bar{S}$, is known and the exponential growth constant k, is determined (usually by ln 2/doubling time). The same applies to all prior and subsequent steps to S. Each prior flux is greater than $\bar{J}_1$ by the expansion flux of the prior metabolite and each subsequent flux is smaller by the expansion flux of the subsequent metabolite. The expanding steady state is therefore constrained by different relations than the stationary ones. In particular, there is no single pathway flux and the earlier the enzyme, the larger the flux it carries.

It will be recognized that Scheme (5) is a special case of a divided pathway shown in Scheme 2. Each expansion flux in a pathway will therefore have a negative control coefficient for fluxes distal to its pool. Although these are virtual coefficients in the sense that no enzymes are involved they must be taken into account when considering the control distribution. Control coefficients must now specify not only the enzyme, but also the flux in the segment of the pathway. The summation property still applies. When we consider any flux in the whole system, then

$$\sum_{i=1}^{n} C_{E_i}^{J_m} + \sum_{j=1}^{n} C_{S_j}^{J_m} = 1$$

The number of control coefficients is therefore the total number of enzymes plus the total number of pools. In a simple pathway all enzymes and the pools distal to the measured flux will have positive coefficients while all proximal pools will have negative ones

$$\sum_{i=1}^{n} C_{E_i}^{J_m} + \sum_{i=1}^{r} C_{X_{p_i}}^{J_m} - \sum_{j=s}^{n} C_{X_{p_j}}^{J_m} = 1$$

where pools 1 to r are distal to J_m and pools s to n are proximal.

Since each expansion flux is $= S_i \cdot k$, modulation of all the expansion fluxes simultaneously would result if the growth rate were to change by, for example, some inhibition of the protein synthesis machinery affecting growth rate.

One consequence of the existence of negative expansion coefficients is that the sum of all the (positive) enzymic coefficients can be >1. This is particularly the case if the steady-state pools of the negative coefficients are large. (It is a corollary of the existence of expansion fluxes that there is a cost of maintaining the steady state in growing systems, unlike stationary systems where, once the steady state has been established, all inputs contribute to the metabolic output.)

The existence of expansion fluxes has been demonstrated[9,24] experimentally in growing *Neurospora* mycelium by measuring the input and output of part of the arginine pathway. The deficit in the output was totally accounted for by calculating the missing expansion fluxes.

VI. CONCLUSION

In this chapter, I have sketched out the general approach of control analysis and some of its applications. It is essentially a method. It suggests what questions to ask and what experiments to do to answer them. The algebraic treatment is quite simple but has yielded a number of useful theorems (the properties) which illuminate the behavior of the system and the connections between the parts. It has also generated relationships useful to the experimenter. The approach will continue to produce further theorems of more complex groups of reactions, such as moiety conserved cycles,[25] enzyme cascades, and organelles. It provides the link between enzymology and physiology. Each species will present its special problems which will have to be solved, but it is hoped that control analysis will give direction to the investigation of questions which continue to puzzle us.

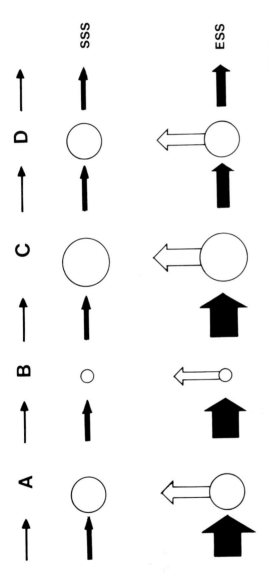

FIGURE 5.   Pools and fluxes in steady-state systems.   The thickness of the arrows represents the magnitude of the fluxes and the diameter of the intermediate pools. For any given pool pattern, the fluxes are different in the two systems. Metabolic fluxes are shown as black arrows, and expansion fluxes as open arrows.   Abbreviations: SSS, stationary steady state; ESS, expanding steady state.

REFERENCES

1. **Kacser, H. and Burns, J. A.,** The control of flux, in *Rate Control of Biological Processes,* Davies, D. D., Ed., Cambridge University Press, London, 1973, 65.
2. **Heinrich, R. and Rapoport, T. A.,** A linear steady state treatment of enzymatic chains, *Eur. J. Biochem.,* 42, 97, 1974.
3. **Heinrich, R. and Rapoport, T. A.,** Mathematical analysis of multienzyme systems. II. Steady state and transient control, *Biosystems,* 7, 130, 1975.
4. **Kacser, H. and Burns, J. A.,** Molecular democracy: who shares the controls?, *Biochem. Soc. Trans.,* 7, 1149, 1979.
5. **Kacser, H. and Burns, J. A.,** The molecular basis of dominance, *Genetics,* 97, 639, 1981.
6. **Kacser, H.,** The control of enzyme systems *in vivo:* elasticity analysis of the steady state, *Biochem. Soc. Trans.,* a11, 35, 1983.
7. **Groen, A. K., van der Meer, R., Westerhoff, H. V., Wanders, R. J. A., Ackerbroom, T. P. M., and Tager, J. M.,** Control of metabolic fluxes, in *Metabolic Compartmentation,* Sies, H., Ed., Academic Press, New York, 1982, 9.
8. **Groen, A. K., Wanders, R. J. A., Westerhoff, H. V., van der Meer, R., and Tager, J. M.,** Quantification of the contribution of various steps to the control of mitochondrial respiration, *J. Biol. Chem.,* 257, 2754, 1982.
9. **Flint, H. J., Tateson, R. W., Barthelmess, I. B., Porteous, D. J., Donachie, W. D., and Kacser, H.,** Control of the flux in the arginine pathway of *Neurospora crassa:* modulations of enzyme activity and concentration, *Biochem. J.,* 200, 231, 1981.
10. **Westerhoff, H. V., Groen, A. K., and Wanders, R. J. A.,** Modern theories of metabolic control and their application, *Biosci. Rep.,* 4, 1, 1984.
11. **Burns, J. A., Cornish-Bowden, A., Groen, A. K., Heinrich, R., Kacser, H., Porteous, J. W., Rapoport, S. M., Rapoport, T. A., Stucki, J. W., Tager, J. M., Wanters, R. J. A., and Westerhoff, H. V.,** Control analysis of metabolic systems, *Trends Biochem. Sci.,* 10, 16, 1985.
12. **Fell, D. A. and Sauro, H. M.,** Metabolic control and its analysis. Additional relationships between elasticities and control coefficients, *Eur. J. Biochem.,* 148, 555, 1985.
13. **Westerhoff, H. V. and Chen, Y.,** How do enzyme activities control metabolite concentrations? An additional theorem in the theory of metabolic control, *Eur. J. Biochem.,* 142, 425, 1984.
14. **Tager, J. M., Groen, A. K., Wanders, R. J. A., Duzynski, J., Westerhoff, H. V., and Vervooorn, R. C.,** Control of mitochondrial respiration, *Biochem. Soc. Trans.,* 11, 40, 1983.
15. **Wanders, R. J. A., Groen, A. K., Van Roermund, C. W. T., and Tager, J. M.,** Factors determining the relative contribution of the adenine nucleotide translator and the ADP regenerating system to the control of oxidative phosphorylation in selected rat liver mitochondria, *J. Theor. Biol.,* 56, 51, 1984.
16. **McMinn, C. L. and Ottoway, J. H.,** On the control of enzyme pathways, *J. Theor. Biol.,* 56, 51, 1976.
17. **Salter, M., Knowles, R. G., and Pogson, C. I.,** Quantitation of the importance of individual steps in the control of aromatic amino acid metabolism, *Biochem. J.,* 234, 635, 1986.
18. **Barthelmess, I. B., Curtis, C. F., and Kacser, H.,** Control of the flux to arginine in *Neurospora crassa:* derepression of the last three enzymes of the arginine pathway, *J. Mol. Biol.,* 87, 303, 1974.
19. **Burns, J. and Kacser, H.,** Allosteric repression: an analysis, *J. Theor. Biol.,* 68, 199, 1977.
20. **Stuart, F., Porteous, D. J., Flint, H. J., and Kacser, H.,** Control of the flux in the arginine pathway of *Neurospora crassa:* Effects of co-ordinate changes of enzyme concentration, *J. Gen. Microbiol.,* 132, 1159, 1986.
21. **Niederberger, P.,** Personal communication.
22. **Kradolfer, P., Niederberger, P., and Hutter, R.,** Tryptophan degradation in *Saccharomyces cerevisiae:* characterisation of two aromatic aminotransferases, *Arch. Microbiol.,* 133, 242, 1982.
23. **Miozzari, G., Niederberger, P., and Hutter, R.,** Tryptophan biosynthesis in *Saccharomyces cerevisia:* control of the flux through the pathway, *J. Bacteriol.,* 134, 48, 1978.
24. **Flint, H. J., Porteous, D. J., and Kacser, H.,** Control of flux in the arginine pathway of *Neurospora crassa:* the flux from citrulline to arginine, *Biochem. J.,* 190, 1, 1980.
25. **Hofmeyer, J. S., Kacser, H., and van der Meer, K. J.,** Metabolic control analysis of moiety conserved cycles, *Eur. J. Biochem.,* 155, 631, 1986.
26. **Reichert, U.,** Gendosenwirking in einen *ad-2* Mutantensystem bei *Saccharomyces, Zentrabl. Bakteriol.,* 205, 63, 1967.
27. **Dean, A. M., Dykhuizen, D. E., and Hartl, D. L.,** Fitness as a function of β-galactosidase activity in *Escherichia coli, Genet. Res. Camb.,* 48, 1986.
28. **Gianelli, F. and Pawsey, G. A.,** DNA repair synthesis in human heterokaryons. III. The rapid and slow complementing varieties of *Xeroderma pigmentosum, J. Cell Sci.,* 20, 207, 1976.
29. **Middleton, R. J. and Kacser, H.,** Enzyme variation, metabolic flux and fitness: alcohol dehydrogenase in *Drosophila melanogaster, Genetics,* 105, 633, 1983.

30. **Torres, N. V., Mateo, F., Melendez-Hevia, E., and Kacser, H.,** Kinetics of metabolic pathways. A system *in vitro* to study the control of flux, *Biochem. J.,* 234, 169, 1986.
31. **Walsh, K. and Koshland, D. E.,** Characterization of rate controlling steps *in vivo* by use of an adjustable expression vector, *Proc. Natl. Acad. Sci. U.S.A.,* 82, 3577, 1985.
32. **Niederberger, P. et al.,** Personal communication.
33. **Woodrow, I. E.,** Control of the role of photosynthetic carbon dioxide fixation, *Biochim. Biophys. Acta,* 851, 181, 1986.

Chapter 2

A THERMODYNAMIC VIEW OF BACTERIAL GROWTH

K. van Dam, M. M. Mulder, J. Teixeira de Mattos, and H. V. Westerhoff

TABLE OF CONTENTS

I. INTRODUCTION

Bacteria can be considered as catalysts that are able to convert materials into products; among these are identical copies of themselves. One may be interested in one or more of these products, in which case it is desirable to formulate the conditions under which the efficiency of transformation of substrates into these products is maximal. To this end, a number of model descriptions of bacterial growth and product formation have been formulated. It is the purpose of this review to consider some general aspects of these models, especially in the light of the boundary conditions that thermodynamic laws impose on the chemical reactions involved.

If one surveys what is known about microbial physiology, one almost immediately notices a strong dichotomy. On the one hand, much is known (empirically) about the relationships between growth rate and rate of substrate consumption for different microorganisms growing on various substrates. On the other hand, the biochemistry and molecular biology of the microorganisms are known in considerable detail. However, between these two well-understood aspects of microbial growth there is a domain where our understanding is far less complete: this concerns the question of how the growth kinetics of microorganisms are related to their biochemistry.

Interest in this question is strong, as is easily understood. For one thing, there is the conviction that growth kinetics should be completely determined by (changes in) the metabolism (i.e., the biochemistry) of the microbial cell. Also, there is the promise that if one could succeed in elucidating the relationship between the cell's biochemistry and its growth kinetics, one could immediately proceed and implement all existing biochemical knowledge to steering and regulating microbial growth and product formation. Furthermore, one would like to compare the efficiency of biochemical systems to that of manmade machines, to assess the economic potential of biotechnology.

In recent years, new approaches to the modeling of microbial growth have appeared, which aimed specifically at closing the gap between biochemistry and microbial growth kinetics. In this chapter we discuss the position of these approaches relative to other models of microbial growth. We conclude that the new approaches allow us to understand growth kinetics specifically in terms of microbial free-energy metabolism.

A limitation, however, in all descriptions is the assumption that enzyme activities do not vary significantly with growth rate. We note that for further development of this field it is most urgent (1) to establish to what extent enzyme activities vary with growth rate, (2) to analyze to what extent the varying enzyme activities control growth, and (3) to develop theoretical growth models in such a manner that they take varying enzyme activities into account.

II. UNDERSTANDING GROWTH KINETICS AT THE PHENOMENOLOGICAL LEVEL

In this section we shall review models of microbial growth that do not relate growth kinetics to cellular biochemistry, but are merely descriptive (phenomenological).

The original concept of yield, the amount of cells (usually grams) arising from substrate (usually grams or moles), was purely phenomenological and followed from the early work of Monod,[1] who noticed a constant relationship between growth and carbon substrate consumption for various carbon substrates. A proportional relationship between the specific increase in bacterial density (x) and the specific substrate (s) consumption rate was observed:

$$1/x \cdot \frac{dx}{dt} = -Y_s \cdot 1/x \cdot \frac{ds}{dt} \qquad (1)$$

where the proportionality factor Y_s is defined as the growth yield.

In the microbiological literature the specific growth rate is usually denoted by μ and the specific consumption rate by q_s. We prefer to use the nonequilibrium thermodynamic symbols. The definitions are

$$\mu = J_a = 1/x \cdot dx/dt \tag{2}$$

and

$$q_s = J_s = 1/x \cdot ds/dt \tag{3}$$

Equation 1 can be written as:

$$J_s = -J_a/Y_s \quad \text{or} \quad Y_s = -J_a/J_s \tag{4}$$

The growth parameter Y_s can be easily determined with reasonable accuracy and this has been done for many organisms and many carbon substrates.[2-10]

In batch cultures the yield on a given substrate appears to be fairly constant, i.e., rather independent of the microorganism actually growing.[2,5,9-16] When different substrates were used as carbon source for growth, the yield per mole carbon varied with the substrate. Here, the first fruits of the application of biochemistry to microbial growth kinetics could be picked: it was realized that the growth substrate was not only needed to provide carbon atoms for the biomass generated, but also to provide the free energy needed for the thermodynamically uphill process of growth.

ATP was recognized as the free-energy currency in the living cell and, reading the number of ATP molecules (n_p^c) produced in catabolism from biochemical maps, the amount of ATP produced in catabolism was calculated as:[17-22]

$$J_p^c = n_p^c \cdot J_c \tag{5}$$

Here J_c refers to the amount of substrate catabolized (expressed in C-moles, moles of substrate times number of carbon atoms in that substrate), and as such, is generally smaller than the total rate of substrate consumption, J_s. In fact, for aerobic growth, J_c is more directly related to oxygen consumption.

With the originally tacit assumption that the amount of ATP (n_p^a) needed to make a progenitor cell from low molecular weight carbon substrates would be independent of the carbon substrate employed, one could now understand why different carbon substrates would lead to different growth yields. First, the amount of biomass produced per ATP used for growth (anabolism), when expressed in C-moles, could be written as:

$$J_a = J_p^a/n_p^a \tag{6}$$

Here, J_p^a is the rate at which ATP is used for growth. Assuming that all ATP used in catabolism would be available for growth, one could equate J_p^a to $-J_p^c$, so that:

$$-J_a = n_p^c \cdot J_c/n_p^a \tag{7}$$

where $-J_a$ is the specific growth rate, defined in Equation 2. With different substrates for catabolism, n_p^c was allowed to vary, but $-J_a/n_p^c \cdot J_c$ was expected to be constant. This parameter was called Y_{ATP}:

$$Y_{ATP} \stackrel{d\,e\,f}{=\!=\!=} -J_a/(n_p^c \cdot J_c) \tag{8}$$

When all ATP produced in catabolism is used for growth, then combining the last two equations:

$$Y_{ATP} = 1/n_p^a \qquad (9)$$

It must be re-emphasized that the fluxes J are experimentally determined, but that the values of Y_{ATP} are calculated on the basis of assumed constants (n_p^c or n_p^a).

Using the values of n_p^c as they can be read from biochemical textbooks, many investigators have determined Y_{ATP} values, both from experiments in batch and in chemostat cultures. It was originally thought that Y_{ATP} was a universal constant, since the determined value was 10.5 (gram cells per mole ATP) with a variety of microorganisms growing on various carbon sources. Differences in $Y_{glucose}$ between two organisms (e.g., *Leuconostoc mesenteroides* and *Streptococcus faecalis*), grown anaerobically on glucose, could be attributed to the amount of ATP that the respective organisms could generate from the energy source (1 and 2 mol ATP per mole glucose fermented, respectively). Further data concerning the quantitative energetic aspects of microbial growth have been reported by Bauchop and Elsden[2] and Rosenberger and Elsden.[9]

While these data seemed to indicate that Y_{ATP} was a constant, they suffered from the limitation that they were obtained in cultures where the microorganisms were allowed to grow at their maximum growth rate. A different picture emerged when yield studies were performed in chemostat cultures with organisms growing at submaximal growth rates. It turned out that there is no proportional relationship between substrate consumption rate and growth rate: even under conditions where the growth rate is zero there is still a definite rate of catabolic substrate consumption.[8,10,15,23-26] This implies that the yield varies with growth rate. To account for this, growth-rate independent substrate utilization was introduced by Pirt:[23]

$$J_s^{total} = J_s^{growth} + J_s^{J_a=0} \qquad (10)$$

and, consequently, Equation 4 has to be extended to:

$$J_s^{total} = -\left(\frac{1}{Y_s^{max}}\right) \cdot J_a + J_s^{J_a=0} \qquad (11)$$

For growth on a substrate that serves both as energy and carbon source, the fraction of J_s that is needed for biomass formation has to be subtracted in order to calculate the rate of catabolism properly. To this end we return to J_c, i.e., the part of the substrate consumption used in catabolism. Expressing J_c, $-J_a$, and J_s in C-moles, and using the conservation property of chemical elements, one finds:

$$J_s = -J_a + J_c \qquad (12)$$

Consequently, Equation 11 can be translated as:

$$-J_a + J_c = \left(\frac{1}{Y_s^{max}}\right) \cdot (-J_a) + J_s^{J_a=0} \qquad (13)$$

$$J_c = \left(\frac{1}{Y_s^{max}} - 1\right) \cdot (-J_a) + J_s^{J_a=0} \qquad (14)$$

This equation shows that

$$J_c^{J_a=0} = J_s^{J_a=0} \qquad (15)$$

For Y_{ATP} this implies:

$$\frac{1}{Y_{ATP}} = \frac{1}{Y_{ATP}^{max}} + \frac{n_p^c \cdot J_c^{J_a=0}}{-J_a} \tag{16}$$

with

$$Y_{ATP}^{max} = \frac{Y_s^{max}}{1 - Y_s^{max}} \Big/ n_p^c \tag{17}$$

Clearly, Y_{ATP} would depend on the growth rate. Equation 17 would suggest, however, that a double reciprocal plot of Y_{ATP} vs. the growth rate $-J_a$ would yield the Y_{ATP}^{max} as the intercept with the ordinate; thus, one may be tempted to conclude that by calculating J_{ATP} values for cultures growing at different dilution rates in a chemostat, one could assess both the so-called "maintenance" rate of ATP generation and the theoretical yield on ATP ($Y_{ATP}^{max. theor.}$), i.e., the Y_{ATP} corrected for processes not associated with growth.

The simplest determination of Y_{ATP} (assuming a specified value for the stoichiometric number n_p^c) seems possible with purely fermentative microorganisms such as homolactic *Streptococci*. However, even in this case one has to be cautious, since the assumption that 1 mol ATP is produced per mole lactate formed from glucose may not be valid under all conditions. The efflux of the lactate generated from the cells may be accompanied by proton efflux and, thereby, significantly increase the value of n_p^c.[27-30]

On the other hand, it has been shown in *Klebsiella aerogenes* that there exists (under certain growth conditions) a pathway that results in the actual consumption of 1 mol ATP per mole lactate formed from glucose.[31] Furthermore, it has been shown that elevated concentrations of many fermentation products, especially weak acids such as acetic and propionic acid, lower the yield; it has been suggested that this is a consequence of the uncoupling of energy generation from catabolism, either by interference of the product with the proton translocating mechanism or by the activation of a futile cycle.[32-34]

Thus, the overall value of n_p^c under anaerobic conditions may vary according to the species of microorganism and the growth conditions from a value >1 ATP/lactate to negative values.

Matters become even more complicated when one has to assign a value to the stoichiometry of ATP synthesis for organisms growing under conditions where energy is generated by oxidative phosphorylation. In this case, ATP synthesis by substrate-level phosphorylation contributes usually only a small fraction of the total energy metabolism and, therefore, the most important parameter in calculating J_{ATP} is the specific rate of reduction of the electron acceptor, say J_{O_2} or $J_{NO_3^-}$. Here, the crux of the matter resides in the assumption one has to make concerning the ATP/acceptor ratio. It is beyond the scope of this review to go into detail concerning the different opinions with regard to this ratio. The point we want to make is that whereas there are already uncertainties about the values of these stoichiometries, we can be certain that the stoichiometries are not true constants (independent of the growth conditions). For instance, it is known that in the yeast *Candida utilis* the presence of site 1 for energy conservation is dependent on the medium composition.[35] Similarly, the P/O ratio of *Paracoccus denitrificans*, grown in chemostat culture, was found to depend not only on the limitation applied, but also on the carbon substrate used.[36] As a last example, it is known that nitrogen fixing cultures of *Azotobacter vinelandii* contain a branched electron transfer chain; only one of these branches has phosphorylating capacity, leading to a highly variable P/O ratio in vivo.[37]

In conclusion, the assessment of n_p^c by considering textbook biochemical pathways is a very risky undertaking and caution has to be exerted, even when seemingly simple cases are studied. Moreover, it should be realized that any value of a stoichiometry that has been found acceptable for one condition has to be reconsidered for any condition. On the other hand, the value of the present approach is that experimental finding of a constant stoichiometry under different conditions is strongly indicative of the absence of any nonconsidered side reactions.

One could also start the determination of Y_{ATP}^{max} by assuming that n_p^a is a constant. Such a Y_{ATP}^{max} would be called $Y_{ATP}^{max. \ theor.}$ and (see Equation 9) would equal $1/n_p^a$. If we knew all biosynthetic pathways in the cell with the concomitant stoichiometric constants for the individual sequences, we could calculate the number of moles ATP required to synthesize 1 C-mol of cells. In fact, we then consider the cell as a collection of end products. This important work has been done by several investigators;[17,19-21] however, it turns out that $Y_{ATP}^{max. \ theor.}$ values calculated in this manner are almost invariably much higher than Y_{ATP}^{max} values determined on the basis of the assumption of constant n_p^c. This finding has been explained by most workers by stating either that the coupling between catabolism and ATP synthesis is less than was assumed (i.e., n_p^c is overestimated) or that the ATP requirement for synthesis of cell material is incorrect (i.e., n_p^a is underestimated). Both explanations illustrate the fundamental failure of defining a complete set of reactions, describing the total cellular metabolism.

It has been pointed out that there are probably two major errors in the approach used to calculate n_p^a. In the first place, since a cell is considered to be the sum of its constituents, the turnover of these constituents, which can be seen as a special form of a futile cycle, is neglected. Because, for instance, protein synthesis has a relatively high ATP requirement and protein turnover may be considerable, this approximation may not be justified. One of the urgently needed pieces of information is what fraction of ATP turnover is related to protein turnover. Secondly, it is very difficult to make reliable assumptions concerning the relative amount of ATP that is required for transport processes during growth. The ATP expenditure for uptake of many organic nutrients is well established, although not always unequivocally, but a major problem arises in the ATP requirement for uptake and retention of inorganic cations or, vice versa, their extrusion.[38,39] For instance, potassium is an important cation for most bacterial species and the internal potassium pool may be as high as 150 mM in rapidly growing Gram positive cells.[40] The internal concentration is highly dependent on the growth rate,[41] and as it is known that potassium may leak from the cell, one can easily envisage potassium cycling across the cell membrane with a rate that is determined by the external potassium concentration and by the growth rate.[42] Thus, a condition-dependent futile cycle is present. Similar arguments may hold for other cations as well as for the export of end products of metabolism from the cell.

The considerations above demonstrate that the approach exemplified by Equation 10 is inadequate. In fact, this approach just matches substrate consumption and growth rate at zero growth rate by adding a constant, $J_s^{J_a \ = \ 0}$. This term, called maintenance, serves to describe the incomplete coupling between growth and substrate consumption. Yet, biochemistry suggests that incomplete or variable coupling can reside in variations in n_p^a and n_p^c, or even in simple hydrolysis of ATP, and it is not *a priori* clear whether the effects of such modes of uncoupling can be represented by the term $J_s^{J_a \ = \ 0}$.

Neijssel and Tempest[24] stressed that the matching term $J_s^{J_a \ = \ 0}$ might also vary with growth rate. They proposed a more general equation:

$$J_s^{total} = \frac{1}{Y_s^{max. \ theor.}} \cdot (-J_a) + \alpha \cdot (-J_a) + J_s^{J_a = 0} \qquad (18)$$

in which α is a constant, i.e., they allowed the maintenance term to vary linearly with growth rate. In fact, even this is an assumption, i.e., the completely general equation would be

$$J_s^{total} = \frac{1}{Y_s^{max. \; theor.}} \cdot (-J_a) + f(-J_a) \qquad (19)$$

The experimental evidence for the fact that Equation 10 with constant $J_s^{J_a \, = \, 0}$ fails to describe microbial growth is threefold: (1) Y_{ATP}^{max} determined according to Equation 18 depends on what limits growth,[24] (2) Y_{ATP}^{max} is not even close to $1/n_p^a = Y_{ATP}^{max. \; theor.}$, and (3) in systems where growth is artificially uncoupled from substrate consumption, the effects of uncoupler cannot be described solely through effects on $J_s^{J_a \, = \, 0}$.

Of course, Equations 18 or 19 allow us to account for these three pieces of experimental evidence. However, with these equations we lose more than we gain; the only "understanding" we are left with is that growth rate might vary linearly with the rate of total substrate consumption. Neither the slope nor the intersection point with the axes of the relationship bear any explicit relation to the biochemical events in the growing cells.

Thus, one ends up in the frustrating situation where, on the one hand one has a (satisfactory)[24] description of microbial growth kinetics (see Equation 18), and on the other hand one has several potential biochemical mechanisms for the uncoupling of growth from catabolism, but no way to relate the two sets of information. Of course, some progress has been made since the formulation of the first growth equation (Equation 1) by Monod; the incompleteness of the description has been reduced to the part concerning the incomplete coupling of growth to catabolism. Yet, it is essentially this part that is presently of the greatest interest, for several reasons:

1. Having observed that Y_{ATP}^{max} is usually only some 50% of its theoretical maximum, one would like to understand to what extent this implies that microbial growth is inefficient in the thermodynamic sense and whether this has some physiological purpose.
2. The stoichiometries of coupling catabolism and growth to ATP synthesis and breakdown are often inescapable quantities; it is relatively difficult to engineer alternative metabolic pathways with increased n_p^c and decreased n_p^a (this would involve the genetic manipulation of several enzymes at a time). It would seem more feasible to modify enzymes or conditions such that coupling at the biochemical level would become more complete (e.g., by making membranes less permeable to protons, or by increasing the tightness of coupling in enzymes). This might then increase growth yields.
3. If indeed uncoupling serves a physiological function, then it might, in certain cases, be useful to engineer cells such that they are increasingly *un*coupled (e.g., for waste water treatment).

Lately, a number of microbial growth models have appeared that deal explicitly with the implications of incomplete coupling for microbial growth. These make use of recent developments in (nonequilibrium) thermodynamics.[22,43] We shall discuss these at some length here, because one of them suggests why microbial growth is ill-coupled and another allows the explicit analysis of the implications of different modes of uncoupling for the overall kinetics of microbial growth.

III. A PHENOMENOLOGICAL DESCRIPTION OF THE THERMODYNAMICS OF MICROBIAL GROWTH: WHY IS COUPLING INCOMPLETE?

A growing bacterium may be considered as an energy transducer that consumes substrates

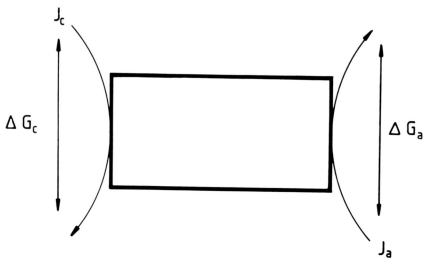

FIGURE 1.   The black box model of microbial growth. The catabolic input is defined by its flux (J_c) and free-energy difference (ΔG_c) and the anabolic output similarly by J_a and ΔG_a.

and produces biomass and other products (Figure 1). It is important to note that for the sake of description, all cells are together represented by one "super-cell". The contents of the vessel containing cells and medium are essentially divided into two compartments: the biotic and the abiotic compartments.[44] Thermodynamically speaking, the cell is an open system. In such a system a balance equation can be formulated for each extensive quantity (i.e., a quantity of which the magnitude is proportional to the amount of substance present): changes in such a quantity are the result of appearance and disappearance, either through chemical transformations or through transport.[44,45] The simplest case to consider is that of a steady state, a situation that is, for instance, experimentally approximated by a continuous culture in the chemostat. In the steady state there is, by definition, no net change in any of the quantities within the cell. Under those conditions one can write down a simple balance equation for all chemical species:

$$J_{net} = 0 \qquad (20)$$

The formation of any species must be exactly balanced by its consumption during the steady state. Apart from this boundary condition of conservation of chemical species, there is also the thermodynamic boundary condition of conservation of energy. The first law of thermodynamics can be formulated in a steady-state system as the net energy flow into a system must be zero. This energy flow can take the form of exchange of heat or exchange of matter carrying its intrinsic enthalpy.

A further restriction is imposed by the second law of thermodynamics: the entropy production must be positive for any realistic system. The most convenient way to use this second law of thermodynamics in the case of bacterial suspensions is by using Gibbs free energies of the chemical species involved. In such systems the Gibbs free energy is a practical parameter, because they usually operate under constant pressure and constant temperature. The entropy production (σ) can then be written as:[43,44]

$$T \cdot \sigma = \Sigma J_i \cdot \Delta \bar{\mu}_i + \Sigma J_r \cdot (-\Delta G_r) \qquad (21)$$

Entropy production consists of terms related to diffusional flows (subscripts i) and terms related to chemical reactions (subscripts r). The total entropy production must be positive.

Furthermore, in the steady state the balance equation for entropy dictates that the entropy production by the system is exactly matched by the entropy flow out of the system.

Finally, a criterion of minimal entropy production in the steady state has been formulated by Prigogine.[46] This criterion, proven to be valid near equilibrium, implies that if all forces in the system are variable there will be only one possible steady state, namely equilibrium (where all forces have dissipated to zero). However, if one of the forces is kept constant, the system will develop to a state where its conjugated flow is zero. This state has been called static head.

In a first approximation, use is made of the linear equations of near-to-equilibrium thermodynamics of irreversible processes.[22] For the case of the black box with input flux of catabolites and output of anabolites (including biomass), the following simple set of phenomenological equations describes catabolism and anabolism:

$$J_a = L_{aa} \cdot \Delta G_a + L_{ac} \cdot \Delta G_c \tag{22}$$

$$J_c = L_{ac} \cdot \Delta G_a + L_{cc} \cdot \Delta G_c \tag{23}$$

Here ΔG_a is the Gibbs free-energy difference per C-mol between the biomass produced and the substrates for growth. ΔG_c similarly is the Gibbs free-energy difference per C-mole between the catabolic end products and the substrate. The symmetrical double occurrence of the cross coefficient L_{ac} is a consequence of the so-called Onsager reciprocity relation.[47] By elimination of one of the driving forces one can easily derive a relation between the catabolic and anabolic fluxes:

$$J_c = \frac{L_{cc}}{L_{ac}} \cdot J_a + \left(L_{ac} - \frac{L_{cc} \cdot L_{aa}}{L_{ac}} \right) \cdot \Delta G_a \tag{24}$$

Such a linear relationship which we recognize in Equations 11 and 18 has been amply documented experimentally in the literature although it holds only strictly for ΔG_a = constant. It is interesting to note that this relationship quite "naturally" allows for a finite rate of catabolism at zero growth rate, a fact that has been attributed to so-called maintenance metabolism in other descriptions.[25]

The derived relationship (Equation 24) also accommodates other phenomena that have been found experimentally, such as the hyperbolic relation between the growth yield $-J_c/J_a$ and the growth rate $-J_a$. It may be noted that Equation 22 implies that there is a logarithmic dependence of the growth rate (J_a) on the limiting substrate concentration. The latter is incorporated in ΔG_a, X which is a logarithmic function of the concentration. Experimental data show that a plot of growth rate vs. substrate concentration fits such a logarithmic function equally well as the commonly used hyperbolic function, introduced by Monod (see Equation 22).

The black-box model of the bacterium is especially appropriate in considering the thermodynamic efficiency of growth and metabolite production. One may equate the black box with a theoretical energy converter as defined by Kedem and Caplan[48] and further analyzed by Stucki.[49]

The thermodynamic efficiency of such a free energy converter is defined by:

$$\eta = -\frac{J_a \cdot \Delta G_a}{J_c \cdot \Delta G_c} \tag{25}$$

It is only this thermodynamic efficiency that is subject to the precise boundary condition imposed by the second law of thermodynamics. These boundary conditions demand that the

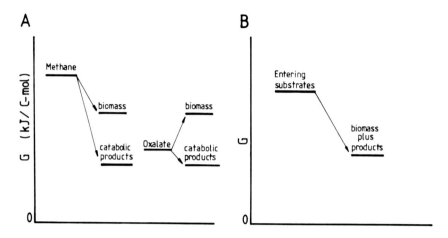

FIGURE 2.   Different views of the energetics of growth. (A) Depending on the free-energy content of the substrate, biomass formation can have a positive or negative ΔG.[43] (B) The free-energy content of the substrates entering the system is always higher than that of biomass and products.[44]

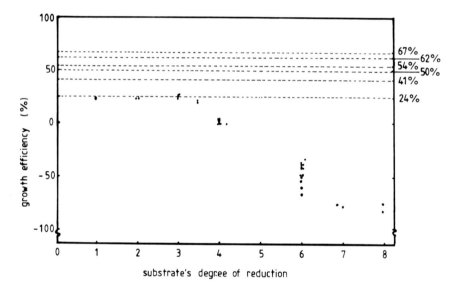

FIGURE 3.   The experimental (●) and theoretical (----) thermodynamic efficiencies of microbial growth as a function of the degree of reduction of the substrate for growth.[50]

efficiency is always less than one, but the value can in principle become negative infinite. Unfortunately, in the literature other definitions of efficiency have been given. For instance, Roels[44] has defined the efficiency as the recovery of enthalpy (in the sense of heat of total combustion) during the process of conversion of catabolites into biomass. This definition is confusing, as it would indicate that it is impossible to have a negative efficiency of growth. The purely thermodynamic definition will allow negative efficiencies. The difference in the two definitions is illustrated in Figure 2.

Combining a number of data from the literature as compiled by Roels,[44] Westerhoff et al.[50] showed that for aerobic growth there is a definite relation between growth efficiency and degree of reduction of the growth substrate. This relation is illustrated in Figure 3.

When the growth substrate is highly reduced, the growth efficiency becomes progressively more negative. The simple biochemical explanation for this finding is that with highly reduced substrates the formation of biomass will be accompanied by synthesis of ATP. In fact, the free energy of the biomass is lower than that of the substrate and, therefore, the overall process is bound to have a negative efficiency. The main concern of the microorganisms will be to dissipate the excess free energy.

At low degree of reduction of the growth substrate the growth efficiency appears to approach a value of around 24%, as indicated in Figure 3. Scanning the literature for data concerning the thermodynamic efficiency of growth, the value of 24% is apparently the highest that is reached in practice.

A thorough theoretical study by Stucki[49] on the optimization of energy converters has led to the conclusion that there may exist several different strategies for optimization. A machine that is constructed to have a maximal power output (in our case maximal free-energy stored as new biomass) will be different from one that has to maintain a maximal force, which, in our case, would mean the generation of the bioenergetically most expensive bacterial progeny.

In principle, changes in output parameters can be brought about in different ways: either the relative magnitudes of the forces or the coupling coefficients may be changed to effect changes in the relative magnitudes of the fluxes in the system. Stucki analyzed how variation in these different parameters would lead to optimal performance of the energy converter with respect to different output parameters.

Translated to the case of bacterial growth[22,43,50] 12 different cases have been recognized, half with a coupling coefficient of exactly 1 and the other half at the coupling coefficient belonging to the optimal efficiency. Each of these cases corresponds to a well-defined thermodynamic efficiency. Four cases that may be realized in biological practice are conditions of maximal: rate of biomass synthesis, power production, "economic biomass synthesis" (rate multiplied by efficiency), or "economic power production" (power production multiplied by efficiency). These cases are accompanied by optimal efficiencies of 24, 41, 54, and 62%, respectively.

It is clear from the data in Figure 3 that the microorganisms growing on rather oxidized substrates approach the efficiency that belongs to energy converters optimized toward maximal rate of biomass synthesis at optimal efficiency. In an evolutionary context these findings may indicate that microorganisms have developed in such a way that a balance has been struck between counteracting tendencies. On the one hand, it is advantageous in the competition between microorganisms to use up available nutrients as quickly as possible, i.e., maximize the rate of consumption of nutrients. On the other hand, to survive in periods of limited nutrient supply, it will be advantageous to produce as much biomass as possible from the available material. We may speculate that the efficiency of growth of 24% on the most demanding substrates is an expression of the optimization strategy followed during evolution.

It should be stressed that the definition of efficiency, formulated by Roels,[44] cannot be used to construct an analysis as described above. Thus, although the enthalpy may be a relatively easily determined parameter, it is not very informative as far as efficiency considerations are concerned.

At this point, it should be noted that consideration of microorganisms as black boxes still discards much of our knowledge of the biochemical machinery inside the cells. Thus, we obtain no insight into the different ways in which the changes in coupling coefficients may have been brought about. Furthermore, the linear approximation for the flux-force relations has been demonstrated to be inappropriate for enzyme-catalyzed (saturable) reactions that are known to take place.[51]

Therefore, it seemed useful to further develop the nonequilibrium description of bacterial growth by opening up the box and considering bacterial metabolism as the sum of a number of independent metabolic elements, each of which has to conform to the laws of thermodynamics and kinetics.

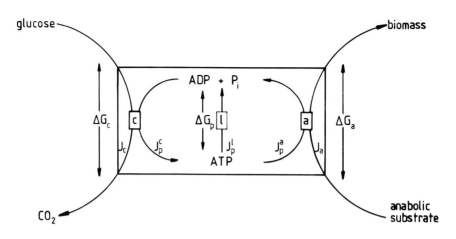

FIGURE 4.   Scheme of microbial growth: the black box opened.

IV. OPENING THE BLACK BOX: MOSAIC NONEQUILIBRIUM THERMODYNAMICS

To illustrate the MNET description of microbial growth[22,43] we use the simple model of Figure 4. In this model the black box has been opened to show its biochemical contents. For simplicity we consider only the coupling reactions between catabolism and ATP synthesis on the one hand and anabolism and ATP utilization on the other. Furthermore, in a real cell, processes occur that lead to hydrolysis of ATP without coupling to either catabolism or anabolism. An example of such a process would be the operation of an ATP driven ion pump (generating a large gradient of a certain ion across the cell membrane) combined with a passive leakage of that ion. The net effect of such a cycle would be the hydrolysis of ATP without any net output. Presence of such ATP "leak" reactions results, of course, in a variable overall coupling between catabolism and anabolism.

It must be realized that the model of Figure 4 is very simplistic and that for a more realistic description many more reactions will have to be included. However, such descriptions would be mathematically more complex and would not help to illustrate our approach.

Thus, bacterial growth is modeled as the sum of three elemental processes which are mutually dependent through the magnitude of the phosphate potential (ΔG_p) only. The first process is the conversion of the growth substrate into catabolic end products with concomitant generation of high free-energy intermediates, in this model ATP. The second process is the conversion of anabolic substrates into biomass, using the high free-energy intermediates generated in catabolism. The third process includes all reactions that lead to changes in the concentration of ATP that are not related to the two former processes; compensation for (passive) ion leakage across the bacterial membrane would be an example.

We stress that in this model for bacterial growth, the only link between anabolism and catabolism considered is the free energy of ATP hydrolysis. It is assumed that there are neither allosteric regulatory links between catabolism and anabolism, nor links through induction or repression of gene expression. For instance, any effect of an increase in concentration of the substrate for catabolism on growth rate is considered to run through an increased ΔG_p, which then increases the free-energy drive for growth (or, if catabolism and anabolism have the substrate in common, through a direct effect on ΔG_a).

For reactions that are not close to equilibrium and, furthermore, catalyzed by enzymes, the usual proportional approximation of nonequilibrium thermodynamics in general cannot be used. It has been shown that a better approximation[43,52] is given by the linear equation:

$$J_i = L_i \cdot (\Delta G_i - \Delta G_i^{\#}) \tag{26}$$

in which the value of $\Delta G_i^{\#}$ depends on the type of reaction and the region of the flow-force relation to be considered. We can apply this more general equation to each of the elementary processes postulated in Figure 4. We note that the activity of the enzyme catalyzing the process i is uniquely represented by its proportionality to L_i. The term $(\Delta G_i - \Delta G_i^{\#})$ serves to indicate that over a significant domain of values for ΔG_i, J_i varies linearly, though not proportionally, with ΔG_i (i.e., $\Delta G_i^{\#}$ is constant, but generally not equal to zero). Outside this domain the variation of J_i with ΔG_i is more complex, but at high and low values of ΔG_i, $(\Delta G_i - \Delta G_i^{\#})$ tends to become constant (the reaction becomes saturated with its substrate or product). For microbial growth this implies that whenever a substrate-product combination is *not* growth limiting this can be described by taking its $(\Delta G - \Delta G^{\#})$ constant.[52] For anabolism we obtain the following equation:

$$J_a = L_a \cdot [(\Delta G_a - \Delta G_a^{\#}) + n_p^a \cdot \gamma_p^a \cdot (\Delta G_p - \Delta G_p^{\#})] \tag{27}$$

The rate of anabolism is a function of the free-energy difference of anabolism plus the free-energy difference of ATP hydrolysis, the latter weighted by the coupling stoichiometry n_p^a between the two (and a correction factor γ_p^a that denotes that the two partial reactions may not be in the same region of saturation of the enzymes). L_a is proportional to the activity of the anabolic enzymes, considered as a unit. An analogous equation can be written for catabolism:

$$J_c = L_c[(\Delta G_c - \Delta G_c^{\#}) + n_p^c \cdot \gamma_p^c(\Delta G_p - \Delta G_p^{\#})] \tag{28}$$

The rate of change in the ATP concentration is described by the sum of three reactions: that coupled to catabolism, that coupled to anabolism, and the leak reaction. The latter is simply given by:

$$J_p^l = L_p^l \cdot (\Delta G_p - \Delta G_p^{\#}) \tag{29}$$

The coupled reactions lead to:

$$J_p^a = n_p^a J_a \tag{30}$$

and

$$J_p^c = n_p^c \cdot J_c \tag{31}$$

The net flux through ATP is the sum of these three:

$$J_p = J_p^a + J_p^c + J_p^l \tag{32}$$

Thus, we obtain the following set of equations to describe growth in MNET terms on the basis of the model of Figure 4:

$$J_a = \cdot L_a (\Delta G_a - \Delta G_a^{\#}) + n_p^a \cdot \gamma_p^a \cdot L_a (\Delta G_p - \Delta G_p^{\#}) \tag{33}$$

$$J_p = n_p^a \cdot L_a \cdot (\Delta G_a - \Delta G_a^{\#}) + [n_p^{a2} \cdot \gamma_p^a \cdot L_a + n_p^{c2} \cdot \gamma_p^c \cdot L_c + L_p^l] \cdot$$

$$(\Delta G_p - \Delta G_p^{\#}) + n_p^c \cdot L_c \cdot (\Delta G_c - \Delta G_c^{\#}) \tag{34}$$

$$J_c = n_p^c \cdot \gamma_p^c \cdot (\Delta G_p - \Delta G_p^{\#}) + L_c \cdot L_c (\Delta G_c - \Delta G_c^{\#}) \tag{35}$$

In principle, one could try to measure the flows and forces under different conditions to establish whether this description is satisfactory. However, this is not feasible in practice even with this simplified model. It is more realistic to establish experimentally well-defined steady states and to use derived equations that take the steady-state conditions into account.

One steady state that is quickly reached is that of zero net flux of ATP, since the intracellular pool of ATP will turn over rapidly and, thereby adapt to changes in the driving forces within a minute or so. The establishment of this steady state allows us to eliminate one of the forces, using Equation 20:

$$J_p = 0 \tag{36}$$

The elimination results in two equivalent relations between catabolism and anabolism:

$$J_c = \left(\frac{n_p^a}{n_p^c}\right)\left(1 + \frac{L_p^l}{n_p^{a2} \cdot \gamma_p^a \cdot L_a}\right)(-J_a) + \frac{L_p^l}{n_p^a \cdot n_p^c \cdot \gamma_p^a}(\Delta G_a - \Delta G_a^{\#}) \tag{37}$$

or

$$J_c = \left(\frac{n_p^a}{n_p^c}\right)\left(1 + \frac{L_p^l}{n_p^{c2} \cdot \gamma_p^c \cdot L_c}\right)^{-1}(-J_a) + \frac{L_p^l \cdot L_c}{n_p^{c2} \cdot \gamma_p^c \cdot L_c + L_p^l}(\Delta G_c - \Delta G_c^{\#}) \tag{38}$$

In both equations the rate of catabolism is given by the sum of two terms. The first term relates the rate of catabolism to the the the rate of anabolism. This term is composed of a number of constants that are properties of the enzymes catalyzing the reactions involved. The second term relates the rate of catabolism to the free energy of either catabolism (Equation 38) or anabolism (Equation 37).

To be able to use the above equations in practice one has to realize that in many experimental situations the same substance serves both as catabolic and as anabolic substrate. Since it is only the total consumption of that substance that can be measured, it is more convenient to consider that consumption rate:

$$J_s = J_c + (-J_a) \tag{39}$$

The equations are then transformed to:

$$J_s = \left[1 + \frac{n_p^a}{n_p^c}\left(1 + \frac{L_p^l}{n_p^{a2} \cdot \gamma_p^a \cdot L_a}\right)\right](-J_a) + \frac{L_p^l}{n_p^a \cdot n_p^c \cdot \gamma_p^a}(\Delta G_a - \Delta G_a^{\#}) \tag{40}$$

and

$$J_s = \left[1 + \frac{n_p^a}{n_p^c}\left(1 + \frac{L_p^l}{n_p^{c2} \cdot \gamma_p^c \cdot L_c}\right)^{-1}\right](-J_a) + \frac{L_p^l \cdot L_c}{n_p^{c2} \cdot \gamma_p^c \cdot L_c + L_p^l} \cdot (\Delta G_c - \Delta G_c^{\#}) \tag{41}$$

The literature contains many experiments to which these equations can be applied. We mention here only a few. In the first place, Equations 37 and 38 predict that there is a linear relationship between substrate utilization and growth when there is either true catabolic, anabolic, or dual substrate limitation, a heavily documented statement (see for instance, Reference 21). Furthermore, the rate of substrate consumption will not fall to zero at zero growth rate. This phenomenon has been attributed to so-called "maintenance" energy requirements. It is immediately clear from the equations that the maintenance energy requirement will increase if the leak reaction increases (L_p^l larger), which at the same time clarifies

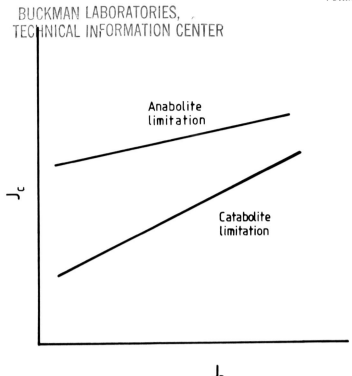

FIGURE 5.   Predicted relation between rate of catabolism and rate of anabolism of microorganisms under different growth limitations.

what the biochemical basis for the maintenance energy requirement could be — the energy requirement for all those processes that consume ATP without coupling to either catabolism or anabolism.

The slope of the line relating the rate of catabolism to the rate of anabolism reflects the "theoretical" stoichiometry of conversion (n_p^a/n_p^c) only if L_p^1 is zero, i.e., when there is no ATP "leakage". If L_p^1 has a positive value, the slope of the line will be larger than the theoretical stoichiometry (in the case of Equation 40). This means that the same increase in growth rate requires a relatively larger increase in substrate utilization or, in other words, there is a growth-rate dependent loss of material conversion.

If one wants to test Equations 37 and 38, they are applicable to different experimental conditions. When the catabolic substrate is limiting for growth, the last term in Equation 37 is constant; this is most easily seen by considering that under those conditions the substrates for anabolism are present in excess and, therefore, the anabolic enzymes are saturated and operating at V_{max}. On the other hand, when the anabolic substrate is limiting for growth, the last term in Equation 38 is constant. Thus, we would apply Equation 37 preferentially to catabolite-limited and Equation 38 to anabolite-limited conditions.

Comparing Equations 37 and 38, we can rationalize the following experimental observations (see Figure 5). The slope of the line relating catabolism to anabolism is steeper when catabolite is limiting (Equation 37) than when anabolite is limiting (Equation 38): in the first case we multiply the theoretical stoichiometry n_p^a/n_p^c by a number larger than one, in the second case, we divide by a number larger than one. Furthermore, the growth-rate independent maintenance energy requirement is larger for anabolite-limited than for catabolite-limited conditions.

In fact, under catabolite-limited conditions, the slope of the plot exceeds the theoretical coupling stoichiometry because of the presence of the ATP leakage reaction. On the other

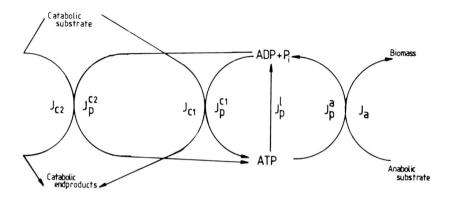

FIGURE 6.   Scheme of a futile cycle in the catabolic branch in metabolism.

hand, under anabolite-limited conditions, the slope in this plot is less than the theoretical value. Thus, it should be possible to estimate the theoretical coupling stoichiometry as lying between these two values.

We should like to introduce a warning about the use of the term "limitation" in general. From the theory of metabolic control as developed by Heinrich and Rapoport[53] and Kacser and Burns[54] and from its applications,[52,55,56] it should be clear that systems are rarely limited or controlled by one factor only. These authors defined a control coefficient as the fractional change in flux through a particular enzymatic step divided by the fractional change in concentration of that enzyme. Thus, if a 1% change in an enzyme causes a 0.5% change in flux, the control coefficient of that enzyme is 0.5. The control coefficient is, however, not a property of the isolated enzyme but of the whole metabolic pathway in which it is embedded. The control coefficients of all enzymes together add up to 1. Thus, it is *a priori* an approximation to describe a process (such as bacterial growth) with one factor as completely limiting.

At this point, we will expand on the question of what mechanisms may exist to effect the ATP leak, introduced in the foregoing model. One possible type of ATP spilling reaction has already been mentioned: the compensation for passive ion leakage. Microorganisms maintain a high gradient of ions that may occur at low concentration in the environment, such as K^+. Also, the electrochemical gradient of H^+ ions, involved in energy transduction, is very high. Although the membrane is impermeable for such ions, there is nevertheless a measurable passive downhill movement which must be exactly matched by ATP (or respiration) driven pumps.

An especially interesting case may exist when two pumps with different stoichiometry exist for the uptake of the same solute.[42] Let us use the example that there are two K^+ uptake systems, one that transports 1 K^+/ATP hydrolyzed and a second that transports 2 K^+/ATP. At equilibrium, the first pump will establish a much higher gradient and the second pump will start to operate in the reverse direction. However, the second pump regenerates only 0.5 ATP/K^+ ion returning across the membrane. Thus, an ATP-utilizing cycle is set up, the activity of which depends on the presence and concentration gradient of K^+. Also, in the main pathways of metabolism in the cell, futile cycles have been postulated. For instance, in the glycolytic pathway the combined action of fructose-6-phosphate kinase and fructose-1,6-diphosphatase will result in ATP hydrolysis (see Reference 57).

To analyze the effect of a futile cycle in the catabolic pathway on the MNET description of bacterial growth, we extend the model of Figure 4 as depicted in Figure 6 (see Reference 22). The catabolic pathway is now represented by two parallel pathways that differ in their stoichiometry of ATP consumption. As a consequence, the catabolic flux must now be described as the sum of two fluxes (we neglect the γ factors):

$$J_{c1} = L_{c1} \cdot \{(\Delta G_a - \Delta G_a^\#) + n_p^{c1}(\Delta G_p - \Delta G_p^\#)\} \tag{42}$$

$$J_{c2} = L_{c2} \cdot \{(\Delta G_a - \Delta G_a^\#) + n_p^{c2}(\Delta G_p - \Delta G_p^\#)\} \tag{43}$$

In the individual pathways we assume strict coupling between fluxes, so that:

$$J_p^{c1} = n_p^{c1} \cdot J_{c1} \tag{44}$$

and $\tag{45}$

$$J_p^{c2} = n_p^{c2} \cdot J_{c2}$$

The total rate of ATP consumption in the catabolic pathway is then:

$$J_p^{ctot} = J_p^{c1} + J_p^{c2} \tag{46}$$

The overall stoichiometry at which ATP is used in catabolism (n_p^c) may be defined as the weighted average:

$$n_p^{c*} \stackrel{d e f}{=} \frac{n_p^{c1} \cdot L_{c1} + n_p^{c1} \cdot L_{c2}}{L_{c1} + L_{c2}} \tag{47}$$

Furthermore

$$L_c^* \stackrel{d e f}{=} L_{c1} + L_{c2} \tag{48}$$

$$L_p^* \stackrel{d e f}{=} \frac{L_{c1} \cdot L_{c2}}{L_{c1} + L_{c2}} \cdot (n_p^{c1} - n_p^{c2})^2 \tag{49}$$

then the following total rate of ATP consumption in catabolism is obtained:

$$J_p^{ctot} = n_p^{c*} \cdot J_c^{tot} + L_p^* \cdot (\Delta G_p - \Delta G_p^\#) \tag{50}$$

The rate of catabolism itself is given by:

$$J_c^{tot} = L_c^* \cdot \{(\Delta G_c - \Delta G_c^\#) + n_p^{c*} \cdot (\Delta G_p - \Delta G_p^\#)\} \tag{51}$$

The general appearance of Equation 51 is similar to that of Equation 28, but not identical: the values of the constants are different. The stoichiometry at which ATP is used in catabolism (n_p^{c*}) is still independent of the forces which drive the pathways. However, the stoichiometry will vary when the relative activity of the pathways is varied (Equation 47).

Obviously, the leakage resulting from the futile cycle disappears if the two pathways have an identical stoichiometry, since in that case L_p^* becomes equal to zero. The leakage term will always be positive if the two pathways differ in their stoichiometry. On the other hand, the leakage term will also vanish if either of the two pathways has a very low activity (L_{c1} or L_{c2} is zero). Evidently, there is a maximum in the value of L_p^*; the value of this maximum can be found by differentiating Equation 49. If L_c^* is kept constant, the maximum value of L_p^* occurs at $L_{c1} = L_{c2}$, i.e., if the conductivity of the two pathways is equal. Thus, a cell can regulate the leakage term by variation of the relative contribution of two pathways with a different stoichiometry of ATP synthesis. As an example we may cite the two possible metabolic routes from dihydroxyacetone phosphate to lactate: one via normal glycolysis and the other via the methylglyoxal pathway. The first pathway leads to the synthesis of 1 ATP/lactate, the second is not coupled to ATP synthesis.[58-60]

The consequences of the modification of the different constants can be seen by deriving equations analogous to Equations 37 and 38. We first eliminate ΔG_p, making use of the $J_p = 0$ steady state condition, which gives:

$$J_a = \left(L_a - \frac{n_p^{a2} \cdot L_a^2}{x} \right)(\Delta G_a - \Delta G_a^{\#}) - \frac{n_p^a \cdot n_p^{c*} \cdot L_a \cdot L_c^*}{x}(\Delta G_c - \Delta G_c^{\#}) \tag{52}$$

and:

$$J_c = -\frac{n_p^a \cdot n_p^{c*} \cdot L_a \cdot L_c^*}{x}(\Delta G_a - \Delta G_a^{\#}) + \left(L_c^* - \frac{n_p^{c*2} \cdot L_c^{*2}}{x} \right)(\Delta G_c - \Delta G_c^{\#}) \tag{53}$$

with

$$x = (L_p^1 + n_p^{a2} \cdot L_a + n_p^{c*2} \cdot L_c^* + L_p^*) \tag{54}$$

These equations can be combined to:

$$J_c = \left(\frac{n_p^a}{n_p^{c*}} \right)\left(1 + \frac{[L_p^1 + L_p^*]}{n_p^{a2} \cdot L_a} \right)(-J_a) + \frac{[L_p^1 + L_p^*]}{n_p^a \cdot n_p^{c*}}(\Delta G_a - \Delta G_a^{\#}) \tag{55}$$

or:

$$J_c = \left(\frac{n_p^a}{n_p^{c*}} \right)\left(1 + \frac{[L_p^1 + L_p^*]}{n_p^{c*2} \cdot L_c^*} \right)^{-1}(-J_a) + \frac{[L_p^1 + L_c^*] \cdot L_c^*}{(n_p^{c*2} \cdot L_c^* + L_p^1 + L_p^*)} \cdot (\Delta G_c - \Delta G_c^{\#}) \tag{56}$$

The modified constants affect both the growth-rate dependent and the growth-rate independent maintenance energy requirement. It may be instructive to take a broad look at the derived equations, by calculating what is the value of the coupling constant q, which quantifies the overall strength with which catabolism and anabolism are coupled. Taking the constants from Equations 52 and 53, we find the following relation:

$$q = \left\{ \left[1 + \frac{(L_p^1 + L_p^*)}{n_p^{a2} \cdot L_a} \right]\left[1 + \frac{(L_p^1 + L_p^*)}{n_p^{c*2} \cdot L_c^*} \right] \right\}^{-1/2} \tag{57}$$

Inspection shows that q will be equal to 1 when both L_p^1 and L_p^* are equal to 0. Changes in L_p^* by simultaneous presence of two pathways will lead to a change in q, which passes through a minimum at a certain ratio of the proportionality constants. Thus, the variations in q that may occur in the microbial cell can be (biochemically) related to changes in the relative contribution of pathways with different stoichiometries. These relative contributions could be varied either through changes in the amount of the enzymes or through variations in the concentration of relevant effectors.

An important difference between the black box approach and the MNET treatment also appears in Equation 57. Evidently, one could reach the same value of q by a change in L_p^1 or by a change in L_p^*. More generally, within certain limits, a specific value of q may arise from many different combinations of the proportionality constants. Although an equal value of q might suggest an equal efficiency, this is not necessarily the case. This will be clear if one considers the general formula describing the efficiency:

$$\eta = \left(q + \sqrt{\frac{L_{aa}}{L_{cc}} \cdot \frac{\Delta G_a}{\Delta G_c}} \right) \left(q + \sqrt{\frac{L_{cc}}{L_{aa}} \cdot \frac{\Delta G_c}{\Delta G_a}} \right)^{-1} \tag{58}$$

If we insert the values of L_{aa} and L_{cc} from Equations 52 and 53, we see that:

$$\frac{L_{aa}}{L_{cc}} = \frac{(L_p^1 + L_p^* + n_p^{c*2} + L_c^{*2})}{(L_p^1 + L_p^* + n_p^{a2} + L_a^2)} \tag{59}$$

Evidently, a change in two of the constants so that q remains constant may still affect the value of η. Thus, the relation between η and the so-called force ratio $(\Delta G_a/\Delta G_c)$ may be different for different combinations of constants, even at the same value of the coupling coefficient q. The only point where the values converge is at the force ratio where the maximal efficiency is reached since there:

$$\eta_{max} = q^2/(1 + \sqrt{1 - q^2})^2 \tag{60}$$

V. VARIATION IN ENZYME ACTIVITIES WITH GROWTH

Up to this point, descriptions of bacterial growth were reviewed in which the bacterium was seen as an energy converter in the first approximation. The metabolic behavior of bacteria, growing under steady-state conditions, could be mathematically described by making use of central concepts of nonequilibrium thermodynamics. As far as specific information about the biochemical make-up of the cell was concerned, an important assumption had to be made, namely that the growth rate (J_a) per se does not influence the overall cell composition (i.e., macromolecular components and enzyme content), nor the specific enzyme activities, because these two parameters determine the numerical value of the constants denoted L in the preceding equations. The approximation that the cell composition is independent of the growth rate is actually implicit in most descriptions of growth, including the Monod model; however, it is without question that the cell composition changes with the growth rate.[61-63] Therefore, a more refined model will have to take this variation into account. As shown in the previous paragraphs, a reasonably satisfactory description can be developed even without including this refinement. A version of the MNET in which variation of enzyme activities is taken into account has not yet appeared. In principle, it is not difficult to see how such a version could be constructed. If, for instance, the activity of the anabolic enzymes is a linear function of the chemical potential of the catabolic substrate, this would lead to a proportionality constant of the form:

$$L_a = L_a^\phi \cdot (1 + \alpha \cdot J_a) \tag{61}$$

in which L_a^ϕ is the value at zero growth rate and α a constant. This could then be used in the MNET equations. A problem met when devising this type of description is the fact that relatively little is known quantitatively about induction and repression. Moreover, L_a is really composed of the activities of a number of enzymes involved in anabolism and the extent to which induction or repression of the activity of one of the enzymes affects the overall coefficient L_a is difficult (although not impossible, following the control theory[53,54,56]) to predict. The recent flourishing of metabolic control theory[56] suggests a possible solution. First, one should determine to what extent the different anabolic enzymes control growth.

Second, one should determine to what extent the ones with significant control are repressed or induced.

These considerations illustrate that any quantitative description of bacterial growth requires not only a detailed knowledge of all net biochemical events (and their respective stoichiometries), but also that information is needed about their relative contributions to overall cellular metabolism. Realizing that these relative contributions in most cases will be dependent on the environment, it can be said again that the main value of this approach will be to show the absence of important side reactions, namely in those cases where the simple description turns out to be satisfactory. The dynamic response of bacterial cells to their environment will usually preclude the simple application of textbook biochemistry to derive values of n_p^a or n_p^c. Nevertheless, the use of known biochemical pathways to describe bacterial growth is in principle an attractive approach because it ultimately aims at connecting all biochemical events at the molecular level. From there on, insight into the energetics of this process could be gained.

The question of whether the practical observation that we obtain fewer cells than expected from the growth substrate is due to impaired energy generation, energy spilling, or poor coupling between ATP hydrolysis and biosynthesis may seem a semantic one from the standpoint of biochemical bookkeeping. However, it brings to light an important difference between the description using yield factors and the MNET approach. The latter has the potential to discriminate between the three main fluxes when their respective stoichiometric constants are changing. By studying the metabolic changes and the various growth parameters, we may experimentally decide with the equations given earlier what role a given catabolic or anabolic pathway plays in the overall metabolism of a bacterial cell.

In order to meet the prerequisite of being a model, i.e., provide us with an insight into the events taking place and their interrelationships, both the MNET and the biochemical descriptions have to introduce simplifications. A very important and questionable assumption in both approaches is that the cell is considered as a unit of constant composition and expressing constant enzyme activities at different growth rates. We know that in general this assumption is not valid. Unfortunately, incorporating the dependency of cell composition on growth rate deprives the biochemical description from most of its value: for each new growth condition the cell composition would have to be determined. Incorporating the variable enzyme activities in the MNET description (e.g., by varying the values of L with growth rate) is not impossible, but will certainly lead to more complex equations.

VI. THE CELL AS THE SUM OF ALL COMPONENTS

In the preceding sections we described how the knowledge concerning the biochemical pathways within bacterial cells could be used to define yield factors. These yield factors give in reality a measure of the relative rate of the different pathways within the cell. However, in most of the work reported in the literature along these lines, little attention has been paid to the actual kinetics of the different pathways. Ideally, one might want to describe the functioning of a cell by the sum of the kinetics of all the enzymes present within that cell. However, such a description is at present not practicable because not all parameters of all enzymes are known and, more seriously, it would require enormous amounts of computer time. Furthermore, the large number of parameters used in such a description would obscure the insight into the crucial steps of metabolism, although such insight might be obtained from a sensitivity analysis.

Therefore, a kinetic description of a bacterial cell usually starts by lumping together groups of metabolic steps, as has been done in the MNET treatment. Depending on the interest of the investigator, emphasis is placed on one or a few selected steps in the kinetic description. The classical example is the Monod model for bacterial growth. It can be viewed as a description of uptake of the limiting substrate by a saturating process (Michaelis-Menten kinetics), while the rest of the metabolism is lumped together as a series of reactions with a rate equal to the rate of uptake of the limiting substrate.

The Monod model has had considerable success in describing the growth of microorganisms as a function of the concentration of the limiting substrate, although it should be noted in passing that surprisingly few actual measurements of the concentration of the limiting substrate have been reported. The model is specifically appropriate for situations of steady-state growth, such as in a chemostat culture. Since the Monod model was so successful, it has been used as the basis for more refined models that give better fits for specific conditions. For instance, Smouse[64] restated the model in a mathematical form that is reminiscent of a Lotka-Volterra model, describing the relation between predator and prey, where the latter feeds on a constant supply of food. This formulation can be easily extended to cases where more than one substrate is limiting growth. An important result of this analysis is that it may be possible under such conditions that multiple genotypes coexist in the steady state, a situation that would not be allowed by the simple Monod model.

An example of the use of a more complete kinetic model of the bacterium is given in the work of Domach et al.[65,66] These authors approached the problem by looking at a single cell and tried to describe its properties as the sum of those of its components. A computer simulation of the kinetics of the enzymes for uptake of substrates, metabolism of the nutrients, duplication and transcription of DNA, synthesis of cell wall components, etc., was set up. Critical factors in the model were the signals that resulted in cell division and the feedback of intracellular components on the uptake of nutrients. By postulating plausible kinetic properties of the different processes, an internally consistent simulation could be constructed. This simulation led to the unanticipated prediction of changes in cell size, cell shape, and timing of chromosome synthesis in response to changes in external glucose limitation. These predictions correlated well with experimental observations.

Although the computer model of Domach et al.[65,66] is satisfactory in some respects, it is already rather complicated and yet leaves a number of important aspects of cellular metabolism unaccounted for. For instance, the largest part of the cellular ATP turnover has to be included in the model in a reaction analogous to the ATP leak reaction used in the MNET treatment. Nevertheless, the model can be considered more satisfactory than simple curve fitting.

Since the model of Domach et al.[65,66] explicitly starts with a single cell, one has to add a factor of cellular variability to it in order to obtain a description of a bacterial population. An example of an approach to assess experimentally the distribution of cells among states is given by the work of Bailey.[67] This author showed that the distribution of cells over the cell cycle depends significantly on the specific growth rate. An analysis of the frequency distribution in different steady states of balanced growth can be used to obtain critical information, required for the single-cell kinetic description.

Another example of a structured model is given by the work of Papageorgakopoulou and Maier.[68] They describe cell growth in terms of two potentially rate-limiting enzyme systems. The model can be applied both to steady-state conditions and to varying conditions, such as occur in batch cultures. The overall type of mathematical simulation is comparable to that in the model of Domach et al.[65]

VII. CONCLUSION AND PROSPECTS

It is clear from this chapter that bacterial growth and functioning can be described at different levels of sophistication, starting from a black box all the way to a complete set of enzymes with their kinetic peculiarities. Which description is the most appropriate will depend on the requirements of the investigator. For practical purposes, one will have to strike a balance between the available information that could potentially be incorporated into the description plus the available computer facilities and the use that will be made of the results. In the context of the use of biotechnological reactors, one would probably be satisfied with a relatively broad model. In investigations concerning the intrinsic properties of microorganisms, more detail should be required, especially on critical steps in the metabolic performance of the cells. A serious drawback of most available models is that they consider the cell as a more or less unchanging entity. It is on this point that the models starting from a complete single cell are superior to all broad models. The latter will compare cells under different growth conditions, assuming that they have the same overall make-up even though it is experimentally known that this is not the case. One of the most urgent tasks in modeling bacterial growth is to also incorporate this knowledge in the descriptions of intermediate complexity.

In our opinion, the MNET model, with its intermediate complexity, is very well suited to analyze the efficiency of bacterial metabolism and, in particular, to pinpoint via relatively few experimental measurements where the quantitatively important "metabolic leaks" occur. Such an analysis will be required in cases where one wants to improve the efficiency of (industrial) microbial processes, for instance, via genetic engineering.

ACKNOWLEDGMENTS

This work was supported in part by The Netherlands Organization for the Advancement of Pure Research (Z.W.O.) under the auspices of The Netherlands Foundation for Chemical Research (S.O.N.).

We wish to thank Dr. P. W. Postma for critical comments on different versions of this manuscript.

REFERENCES

1. **Monod, J.,** *Récherches sur la Croissance des Cultures Bacteriennes,* Hermann, Paris, 1942.
2. **Bauchop, T. and Elsden, S. R.,** The growth of micro-organisms in relation to their energy supply, *J. Gen. Microbiol.,* 23, 457, 1960.
3. **Beck, R. W. and Stugart, L. R.,** Molar growth yields in *Streptococcus faecalis var. liquefaciens, J. Bacteriol.,* 92, 802, 1966.
4. **Brown, W. V. and Collins, E. B.,** End product and fermentation balances for lactic *Streptococci* grown aerobically on low concentrations of glucose, *Appl. Environ. Microbiol.,* 33, 38, 1977.
5. **Crabbendam, P. M., Neijssel, O. M., and Tempest, D. W.,** Metabolic and energetic aspects of the growth of *Clostridium* on glucose in chemostat culture, *Arch. Microbiol.,* in press.
6. **Hadjipetrou, L., Gerrits, J. P., Teulings, F. A. G., and Stouthamer, A. H.,** Relation between energy production and growth of *Aerobacter aerogenes, J. Gen. Microbiol.,* 36, 139, 1964.
7. **Hamilton, I. R., Phipps, P. J., and Ellwood, D. C.,** Effect of growth rate and glucose concentration on the biochemical properties of *Streptococcus mutans Ingbritt* in chemostat culture, *Infect. Immunol.,* 26, 861, 1979.
8. **Neijssel, O. M. and Tempest, D. W.,** The regulation of carbohydrate metabolism in *Klebsiella aerogenes* NCTC 418 organisms, growing in chemostat cultures, *Arch. Microbiol.,* 106, 251, 1976.
9. **Rosenberger, P. F. and Elsden, S. R.,** The yields of *Streptococcus faecalis* grown in continuous culture, *J. Gen. Microbiol.,* 22, 726, 1960.
10. **Teixeira de Mattos, M. J. and Tempest, D. W.,** Metabolic and energetic aspects of the growth of *Klebsiella aerogenes* NCTC 418 on glucose in anaerobic chemostat culture, *Arch. Microbiol.,* 134, 80, 1983.
11. **Buchanan, B. and Pine, L.,** Path of glucose breakdown and cell yields of a facultative anaerobe *Actinomyces naeslundii, J. Gen. Microbiol.,* 46, 225, 1967.
12. **Hernandez, E. and Johnson, M. J.,** Anaerobic growth yields of *Aerobacter cloacae* and *Escherichia coli, J. Bacteriol.,* 94, 991, 1967.

13. **Oxenburgh, M. S. and Snoswell, A. M.,** The use of molar growth yield for the evaluation of energy producing pathways in *Lactobacillus plantarum, J. Bacteriol.,* 89, 913, 1965.

14. **De Vries, W., Kapteyn, W. M. C., Van de Beek, E. G., and Stouthamer, A. H.,** Molar growth yields and fermentation balances of *Lactobacillus casei* L3 in batch cultures and in continuous culture, *J. Gen. Microbiol.,* 63, 333, 1970.

15. **Russel, J. B. and Baldwin, R. L.,** Comparison of maintenance energy expenditures and growth yields among several rumen bacteria grown in continuous culture, *Appl. Environ. Microbiol.,* 37, 537, 1979.

16. **Rattcliffe, H. D., Drozd, J. W., and Pull, A. T.,** Growth energetics of *Rhizobium leguminosarum* in chemostat culture, *J. Gen. Microbiol.,* 129, 1697, 1983.

17. **Hespell, R. B., and Bryant, M. P.,** Efficiency of rumen microbial growth; influence of some theoretical and experimental factors on Y_{ATP}, *J. Anim. Sci.,* 49, 1640, 1979.

18. **Kwaadsteniet de, J. X., Jager, J. C., and Stouthamer, A. H.,** A quantitative description of heterotrophic growth in micro-organisms, *J. Theor. Biol.,* 57, 103, 1976.

19. **Stouthamer, A. H.,** Determination and significance of molar growth yields, in *Methods in Microbiology I,* Norris, J. R. and Ribbons, D. W., Eds., Academic Press, New York, 1969, 629.

20. **Stouthamer, A. H.,** A theoretical study on the amount of ATP required for synthesis of cell material, *Antonie van Leeuwenhoek, J. Microbiol. Serol.,* 39, 545, 1973.

21. **Stouthamer, A. H.,** The search for correlation between theoretical and experimental growth yields, in *International Review of Biochemistry,* Vol. 21, Quayle, J. R., Ed., University Park Press, Baltimore, 1979, 1.

22. **Westerhoff, H. V., Lolkema, J. S., Otto, R., and Hellingwerf, K. J.,** Thermodynamics of bacterial growth. Non-equilibrium thermodynamics of bacterial growth. The phenomenological and the mosaic approach, *Biochim. Biophys. Acta,* 683, 181, 1982.

23. **Pirt, S. J.,** The maintenance energy of bacteria in growing cultures, *Proc. R. Soc. London,* 163, 224, 1965.

24. **Neijssel, O. M. and Tempest, D. W.,** Bioenergetic aspects of aerobic growth of *Klebsiella aerogenes* NCTC 418 in carbon-limited and carbon-sufficient chemostat cultures, *Arch. Microbiol.,* 107, 215, 1976.

25. **Pirt, S. J.,** Maintenance energy; a general model for energy limited and energy sufficient growth, *Arch. Microbiol.,* 133, 300, 1983.

26. **Stouthamer, A. H. and Bettenhausen, C. W.,** Energetic aspects of anaerobic growth of *Aerobacter aerogenes* in complex medium, *Arch. Microbiol.,* 111, 21, 1976.

27. **Ten Brink, B. and Konings, W. N.,** Generation of an electrochemical proton gradient by lactate efflux in *Escherichia coli* membrane vesicles, *Eur. J. Biochem.,* 111, 59, 1980.

28. **Michels, P. A. M., Michels, J. P. J., Boonstra, J., and Konings, W. N.,** Generation of an electrochemical proton gradient by the excretion of metabolic end products, *FEMS Microbiol. Lett.,* 5, 357, 1979.

29. **Otto, R., Langeveen, R., Veldkamp, H., and Konings, W. N.,** Lactate efflux induced electrical potential in membrane vesicles of *Streptococcus cremoris, J. Bacteriol.,* 149, 733, 1982.

30. **Otto, R., Sonnenberg, A. S. M., Veldkamp, H., and Konings, W. N.,** Generation of an electrochemical proton gradient in *Streptococcus cremoris* by lactate efflux, *Proc. Natl. Acad. Sci. U.S.A.,* 77, 5502, 1980.

31. **Teixeira de Mattos, M. J., Streekstra, H., and Tempest, D. W.,** Metabolic uncoupling of substrate level phosphorylation in anaerobic glucose-limited chemostat cultures of *Klebsiella aerogenes* NCTC 418, *Arch. Microbiol.,* 139, 260, 1984.

32. **Teixeira de Mattos, M. J., Plomp, P. J. A. M., Neijssel, O. M., and Tempest, D. W.,** Influence of metabolic end products on the growth efficiency of *Klebsiella aerogenes, Antonie van Leeuwenhoek,* 50, 461, 1984.

33. **Zines, D. O. and Rogers, P. L.,** A chemostat study of ethanol inhibition, *Biotech. Bioeng.,* 13, 293, 1971.

34. **Herrero, A. A., Gomez, R. F., Snedecor, B., Tolman, C. J., and Roberts, M. F.,** Growth inhibition of *Clostridium thermocellum* by carboxylic acids: a mechanism based on uncoupling by weak acids, *Appl. Microbiol. Biotechnol.,* 22, 53, 1985.

35. **Aiking, H., Sterkenburg, A., and Tempest, D. W.,** Influence of specific growth limitation and dilution rate on the phosphorylation efficiency and cytochrome content of mitochondria of *Candida utilis, Arch. Microbiol.,* 113, 65, 1977.

36. **Van Verseveld, H. W. and Stouthamer, A. H.,** Growth yields and the efficiency of oxidative phosphorylation during autotrophic growth of *Paracoccus denitrificans* on methanol and formate, *Arch. Microbiol.,* 118, 21, 1978.

37. **Jones, C. W.,** Aerobic respiratory systems in bacteria, *Symp. Soc. Gen. Microbiol.,* 27, 23, 1977.

38. **Brey, R. N., Beck, J. C., and Rosen, B. P.,** Cation/proton antiport systems in *Escherichia coli, BBRC,* 83, 1588, 1978.

39. **Kashket, E. R.,** Stoichiometry of the H-ATP-ase of *Escherichia coli* cells during anaerobic growth, *FEBS Lett.,* 154, 343, 1983.

40. **Tempest, D. W., Dicks, J. W., and Ellwood, D. C.,** Influence of growth condition on the concentration of potassium in *Bacillus subtilis var. niger* and its possible relationship to cellular ribonucleic acid, teichoic acid and teichuronic acid, *Biochem. J.,* 106, 237, 1968.

41. **Tempest, D. W., Dicks, J. W., and Hunter, J. R.,** The interrelationship between potassium, magnesium and phosphorus in potassium-limited chemostat cultures of *Aerobacter aerogenes, J. Gen. Microbiol.,* 45, 135, 1966.

42. **Van Dam, K. and Mulder, M. M.,** Thermodynamic description of bacterial growth, in *Third European Bioenergetics Conference,* Vol. 3B, Schafer, G., Ed., Cambridge University Press, London, 1985.

43. **Westerhoff, H. V.,** Mosaic Non-Equilibrium Thermodynamics and the Control of Biological Free-Energy Transduction, Ph.D. thesis, University of Amsterdam, 1983.

44. **Roels, J. A.,** *Energetics and Kinetics in Biotechnology,* Elsevier, Amsterdam, 1983.

45. **Katchalsky, A. and Curran, P. F.,** *Nonequilibrium Thermodynamics in Biophysics,* Harvard University Press, Cambridge, 1967.

46. **Prigogine, I.,** *Introduction to Thermodynamics of Irreversible Processes,* Interscience, New York, 1955.

47. **Onsager, L.,** Reciprocal relations in irreversible processes, *Phys. Rev.,* 37, 405, 1931.

48. **Kedem, O. and Caplan, S. R.,** Degree of coupling and its relation to efficiency of energy conversion, *Trans. Faraday Soc.,* 21, 1897, 1965.

49. **Stucki, J. W.,** The thermodynamic buffer enzymes, *Eur. J. Biochem.,* 109, 257, 1980; **Stucki, J. W.,** The optimal efficiency and the economic degrees of coupling of oxidative phosphorylation, *Eur. J. Biochem.,* 109, 269, 1980.

50. **Westerhoff, H. V., Hellingwerf, K. J., and Van Dam, K.,** Thermodynamic efficiency of microbial growth is low, but optimal for maximal growth rate, *Proc. Natl. Acad. Sci. U.S.A.,* 80, 305, 1983.

51. **Van der Meer, R., Westerhoff, H. V., and Van Dam, K.,** Linear relation between rate and thermodynamic force in enzyme-catalyzed reactions, *Biochim. Biophys. Acta,* 591, 488, 1980.

52. **Westerhoff, H. V. and Van Dam, K.,** *Mosaic Nonequilibrium Thermodynamics and Control of Biological Free-Energy Transduction,* Elsevier, Amsterdam, 1987.

53. **Heinrich, R. and Rapoport, T. A.,** A linear steady-state treatment of enzymatic chains. General properties, control and effector strength, *Eur. J. Biochem.,* 42, 89, 1974; **Heinrich, R. and Rapoport, T. A.,** A linear steady-state treatment of enzymatic chains. Critique of the crossover theorem and a general procedure to identify interaction sites with an effector , *Eur. J. Biochem.,* 42, 97, 1974.

54. **Kacser, H. and Burns, J. A.,** in *Rate Control of Biological Processes,* Davies, D. D., Ed., Cambridge University Press, London, 1973, 65.

55. **Groen, A. K.,** Quantification of Control in Studies on Intermediary Metabolism, Ph.D. thesis, University of Amsterdam, 1984.

56. **Westerhoff, H. V., Groen, A. K., and Wanders, R. J. A.,** Modern theories of metabolic control and their applications, *Biosci. Rep.,* 4, 1, 1984.

57. **Otto, R.,** Uncoupling of growth and acid production in *Streptococcus cremoris, Arch. Microbiol.,* 140, 225, 1984.

58. **Cooper, R. A.,** The methylglyoxal bypass of the Embden Meyerhof-Parnas pathway, *Biochem. Soc. Trans.,* 3, 837, 1975.

59. **Cooper, R. A.,** Metabolism of methylglyoxal in microorganisms, *Ann. Rev. Microbiol.,* 38, 49, 1984.

60. **Cooper, R. A. and Anderson, A.,** The formation and catabolism of methylglyoxal during glycolysis in *Escherichia coli, FEBS Lett.,* 11, 273, 1970.

61. **Herbert, D.,** The chemical composition of microorganisms as a function of their environment, *Symp. Soc. Gen. Microbiol.,* 11, 391, 1961.

62. **Harder, A. and Roels, J. A.,** Application of simple structured models in bioengineering, *Adv. Biochem. Eng.,* 21, 55, 1982.

63. **Bijkerk, A. H. E. and Hall, R. J.,** A mechanistic model of the aerobic growth of *Saccharomyces cerevisiae, Biotech. Bioeng.,* 19, 267, 1977.

64. **Smouse, P. E.,** Mathematical models for continuous culture growth dynamics of mixed populations subsisting on a heterogeneous resource base. I. Simple competition, *Theor. Pop. Biol.,* 17, 16, 1980.

65. **Domach, M. M., Leung, S. K., Cahn, R. E., Cocks, G. G., and Shuler, M. L.,** *Biotech. Bioeng.,* 26, 203, 1984.

66. **Shuler, M. L. and Domach, M. M.,** in *Foundations of Biochemical Engineering, Kinetics and Thermodynamics of Biological Systems,* ACS Symp. Ser. No. 207, Blanch, H. W., Papoutsakis, E. T., and Stephanopulos, G., Eds., American Chemical Society, Washington, D.C., 1983, 93.

67. **Bailey, J. E.,** in *Foundations of Biochemical Engineering, Kinetics and Thermodynamics in Biological Systems,* ACS Symp. Ser. No. 207, Blanch, H. W., Papoutsakis, E. T. and Stephanopoulos, G., Eds., American Chemical Society, Washington, D.C., 1983.

68. **Papageorgakopoulou, H. and Maier, W. J.,** A new modeling technique and computer simulation of bacterial growth, *Biotech. Bioeng.,* 26, 275, 1984.

Chapter 3

STEADY-STATE KINETIC ANALYSIS OF CHEMIOSMOTIC PROTON CIRCUITS IN MICROORGANISMS*

Dale Sanders

TABLE OF CONTENTS

---

* The manuscript for this chapter was delivered to the editors in December 1985.

I. INTRODUCTION

A. Organization of Transport Across Microbial Membranes

Transport of most ions and other solutes across the membranes of all cells can be viewed as conforming to a simple pattern. A primary pump is responsible for moving an ion (or ions) thermodynamically uphill across the membrane. The primary pump can be coupled to scalar metabolic reactions via redox compounds or via ATP, or it can be powered independently of metabolism by light (as in the case of the H^+ pump of *Halobacterium*, bacteriorhodopsin). In all cases so far investigated in detail, primary pumps separate charge across membranes. They therefore carry electrical current, and are usually referred to (somewhat imprecisely) as being electrogenic. (An influence of the pump on the membrane electrical potential difference is a normal, though not a necessary, consequence of current flow through the pump.)

The electrochemical gradient generated by the primary pump can be utilized dissipatively to drive four further classes of (secondary) transport process. First, downhill flow of the ion across the membrane may be coupled to synthesis of ATP. Second, coupling can occur between the passive flux of the primary ion and uphill transport of some other solute; this process is known as symport when the fluxes are in the same direction, and as antiport for opposing fluxes. Third, the gradient set up by the primary pump can be used to power large, but transient, ion fluxes. The ions involved can be either those transported by the primary pump (as Na^+ and K^+ in animal cells), or, more usually in nonanimal cells, other ions which flow down an electrical gradient for which the primary pump is nevertheless ultimately responsible. Such fluxes constitute the ionic basis for action potentials, and are also involved in control of membrane potential in some cells. Fourth, passive flux of the primary ion through a molecular "motor" can drive flagellar rotation, and hence bacterial movement.

One class of transport system falls outside this general scheme: group translocation, in which chemical modification of the solute is an integral part of the transport process. In bacteria, group translocation is exemplified by the phosphoenolpyruvate phosphotransferase system.

In microorganisms, the primary pump is normally a H^+ pump (although primary pumping of Na^+ or Cl^- occurs in some special systems[1,2]). A transmembrane electrochemical H^+ gradient, $\Delta\bar{\mu}_H$, is therefore regarded as constituting the central element in coupling primary transport to the secondary processes outlined above. This concept was introduced 25 years ago by Mitchell, and is known as "chemiosmotic coupling".[3,4]

Biochemical approaches have yielded considerable insight into the molecular details of primary and secondary transport systems. Obligately anaerobic heterotrophs (e.g., *Streptococcus*) generate their ATP by substrate level phosphorylation and subsequent hydrolysis of ATP is used to pump H^+ out of the cell.[5] The ATPase is a multisubunit enzyme consisting of discrete H^+ translocator (F_0) and adenine nucleotide-binding (F_1) components.[6] Aerobic bacteria, on the other hand, generate a transmembrane H^+ gradient across the inner membrane with the electron transport chain, and return flow of H^+ is coupled to ATP synthesis. However, again ATP synthesis is accomplished by F_0F_1. F_0F_1 is therefore capable of operating in both primary (H^+ pump) and secondary (H^+ leak) modes.

Eukaryotic microorganisms (e.g., fungi) resemble anaerobic bacteria in that the primary H^+ pump of the plasma membrane is ATP-fueled. However, in terms of its reaction chemistry, the pump exhibits a number of fundamental differences to F_0F_1-type pumps.[7] Most notable among these differences is the formation of a phosphoenzyme intermediate in the hydrolysis cycle of the fungal ATPase. The protein consists of a single polypeptide only.

Although the distinction may now be a little blurred,[8] many transport proteins fall into one of two classes. The so-called "channels" are characterized by their high turnover rates (typically of the order 10^6/sec), their capacity to act only dissipatively, and the simultaneous exposure of binding sites to both sides of the membrane.[9] In microorganisms, channels are suspected of involvement in action potentials, and are also well defined as constituting the pores which exist in the outer membrane of Gram negative bacteria. Because channels conduct ionic currents, electrical techniques are normally used for characterization of channel properties.

"Carriers", on the other hand, exhibit rather low turnover rates (10^2/sec), can act either to generate or to dissipate solute gradients, and behave as though a specific binding site is exposed alternately to each side of the membrane.

B. Why Model Transport?

Mathematical modeling has played an essential role in our understanding of enzyme kinetics, giving clues to the mechanism of catalysis and of inhibitor action. Similarly, for transport kinetic data, modeling can provide fundamental clues to some of the underlying reaction chemistry of carrier systems: are the kinetic effects of ligands related exclusively to binding at transport sites, or are there allosteric effects? Do the electrical and chemical driving forces across a membrane act at single or discrete reaction steps? What is the binding order for transported ligands in cotransport systems? What are the relative or absolute values of rate constants which interconnect various carrier states? What is the mode of action of inhibitors?

In a more practical sense, which will become apparent in Section III, modeling can play an essential role in extracting from the general electrical properties of a membrane the specific behavior of an electrogenic pump. This function of modeling takes on special significance in the case of an electrogenic H^+ pump, where the charge-carrying attributes of the transport system may be the only reliable kinetic indicator of its activity.

C. Conceptual Approaches

There are several different ways of setting out to model transport kinetic data. Often, individual studies will incorporate elements of all the approaches described below.

The first, and least precise, method is to take a model for transport, and ask what general properties it exhibits and whether those properties coincide with observation. This approach is normally performed at a nonquantitative or semiquantitative level, but can give essential insights into the model's algebraic properties.

A second, more quantitative, strategy strives to derive constants in some reaction scheme which is assumed to underlie the process under investigation. Curve-fitting techniques are normally used. One danger with this approach is highlighted by the word "assumed": it is easy to lay heavy emphasis on the reaction constants which emerge from the model without questioning the validity of the basic reaction scheme. Both this and the first approach therefore rely on a systematic application of progressively more complex models until one is found which can adequately describe all the data to hand.

Finally, the data themselves may specify a unique model, or at least exclude alternative classes of models. Constants can then be derived from curve-fitting. A prerequisite for model induction is that the behavior of alternative models is known. Normally, therefore, nonquantitative algebraic analysis of model behavior will have been undertaken first.

D. Transport Systems to be Discussed

In this review, I adopt a rather liberal interpretation of the word "microorganism". Rather than restrict the discussion of energy-coupling membranes simply to the bacteria, I have included some discussion of models based on mitochondrial systems. This departure seems justified in view of the many underlying points of similarity in the mechanisms of operation of energy-coupling membranes.

The primary generator and consumers of the proton motive force in eukaryotic microorganisms will also be discussed. Primary charge separation by light-powered systems (e.g., bacteriorhodopsin; bacteriochlorophyll) is not considered. The sequence of rapid light-induced absorbance changes in these systems has meant that it has been possible with elegant spectroscopic techniques to build up a clear quantitative picture of the reaction cycle of the primary pump.[10,11] This picture owes much to the application of nonsteady-state (relaxation) kinetics, which is a very powerful tool in kinetic modeling.[12] The major aim of this review, however, is to survey the role of inductive modeling of steady-state kinetic data from those systems for which a nonsteady-state approach is not yet feasible. A central conclusion is that there is certainly scope — and in one case, a dire need — for a more quantitative approach, based on defined and physically interpretable models.

II. ENERGY-COUPLING MEMBRANES

A. A Localized Coupling Model
1. Supposed Problems with Delocalized Coupling
The power of the chemiosmotic hypothesis[3] became apparent when it was demonstrated that an imposed proton motive force was indeed capable of driving net ATP synthesis in a variety of organellar or microbial systems.[13-15] Problems have, however, arisen in attempting quantitative application of chemiosmotic theory to energy transduction. One model currently enjoying much support as an alternative to straightforward bulk phase-to-bulk phase coupling (as envisaged in the original formulation of chemiosmotic theory) is the "mosaic coupling" hypothesis.[16] This seeks to reconcile observations on bacterial (respiratory and chromatophore), mitochondrial, and thylakoid membranes which have seemingly fallen outside the predictions of chemiosmotic theory into a cohesive hypothesis for H^+-mediated energy coupling.

Westerhoff et al.[17] have discussed three key lines of evidence which appear to contradict the classical view of delocalized H^+ coupling between $\Delta\bar{\mu}_H$ generators (e.g., respiratory chains) and the primary $\Delta\bar{\mu}_H$ consumer, the F_0F_1 ATP synthase.

First, it is pointed out that neither the rate of electron transport nor the rate of phosphorylation is dependent on the absolute magnitude of $\Delta\bar{\mu}_H$. For example, the rate of ATP synthesis in chromatophores is relatively insensitive to uncouplers in comparison with treatments reducing the rate of electron transport, even though the overall change in $\Delta\bar{\mu}_H$ is about the same in either case. Westerhoff et al.[17] contend that classical chemiosmotic theory predicts that H^+ transport-associated functions should show absolute dependence on $\Delta\bar{\mu}_H$.

Second, for a proton-pumping ATPase at equilibrium,

$$n_H^P = -\Delta G_P/\Delta\bar{\mu}_H, \tag{1}$$

where n_H^P is the H^+/ATP stoichiometry and ΔG_P is the Gibbs free energy for ATP hydrolysis. Evidence is cited[17] that at "static head" (no net ATP synthesis), the ratio defined by Equation 1 is dependent on $\Delta\bar{\mu}_H$ such that at low values of $\Delta\bar{\mu}_H$ estimates of n_H^P can exceed 6. Under more highly energized conditions, n_H^P is usually determined as 2 in these experiments. Conventional chemiosmotic theory contends that the value of n_H^P should be fixed.

Third, mild inhibition of electron transport enhances the efficacy of F_0F_1 inhibitors, and vice versa. It is argued[17] on the basis of these "dual inhibitor" titrations that by partially inhibiting H^+ cycling through either the primary or the secondary pathway, rate-limitation through the other pathway should be lifted, and therefore that inhibitors of the nonrate-limiting pathway should, on the basis of conventional chemiosmotic theory, have reduced efficiency. Analogously, it is argued that the efficiency of electron transport inhibitors on reduction of $\Delta\bar{\mu}_H$ should be greater in state 3 conditions (H^+ flux through ATPase) than in state 4 (no net ATP synthesis, or H^+ flux), though this is not observed to be the case.

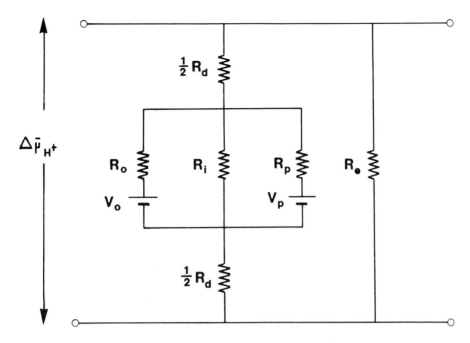

FIGURE 1. Equivalent circuit representation, after Westerhoff et al.,[17] of localized protonic coupling. The coupling unit consists of a proton-pumping electron transport chain (emf, V_o, and internal resistance R_o) and a reversible proton-pumping ATPase (emf, V_p, and internal resistance R_p). Passive proton leakage inside and outside the coupling unit is denoted R_i and R_e, respectively. In normal conditions, with the plane of the membrane horizontal, the cytosolic (or matrix) face will be in the lower half of the figure. $^1/_2R_d$ is the diffusive resistance to proton flow from the membrane surface to the bulk phase. $\Delta\bar{\mu}_H$ is normally measured across the whole equivalent circuit, from bulk phase-to-bulk phase.

Combining these three lines of evidence, Westerhoff et al.[17] concluded that coupling between the primary and secondary systems does not occur via H^+ equilibrated with the bulk phase. Instead, it is proposed that the transport systems are coupled in some more localized manner. An elegant and detailed critique of the experimental evidence for and against delocalized coupling has been published recently,[18] and it is not the intention here to repeat those arguments. Instead, the localized coupling model proposed by Westerhoff et al. is discussed in relation to its assumptions and predictions, and to alternative quantitative models.

2. Linear Equivalent Circuit Model

Westerhoff et al. envisage that the primary and secondary H^+ translocation systems exist in "coupling units". Each coupling unit (present at density n) comprises electron transport, F_oF_1 and proton back-leakage pathways, arranged in parallel in the coupling membrane, with diffusive pathways (enabling equilibration with each bulk phase) arranged in series to this array. In parallel to the whole network are bulk phase-to-bulk phase H^+ leaks (present at density m). The arrangement is represented by a linear electrical equivalent circuit in Figure 1. The conductance and electromotive force (emf) of the electron transport chain are designated $1/R_o$ and V_o, respectively; those of F_oF_1, $1/R_p$ and V_p. The diffusive conductance is $1/R_d$ and the bulk phase-to-bulk phase conductance, $1/R_e$. The scheme was analyzed by application of Kirchoff's laws with respect to its steady state properties.

It is clear that a scheme such as that in Figure 1 is capable of describing some of the apparent deviations of experimental evidence from classical chemiosmotic theory. Thus, assuming that electron transport and phosphorylation are each perfectly coupled to H^+ flux, the rate of either of these processes as a function of $\Delta\bar{\mu}_H$ will depend on whether $\Delta\bar{\mu}_H$ is lowered with ADP or uncouplers.

In the case of electron transport, its overall rate, J_o, as a function of $\Delta\bar{\mu}_H$ when $\Delta\bar{\mu}_H$ is adjusted by addition of ADP and is given as

$$J_o = \frac{n[V_o - \Delta\bar{\mu}_H(1 + mR_d/nR_e)]}{n_H^o FR_o} \tag{2}$$

in which n_H^o is the $H^+/2e^-$ stoichiometry and F is the Faraday constant. In deriving Equation 2, it is assumed that ADP has a direct effect on V_p and R_p. Conversely, for protonophorous uncouplers applied in the condition of static head,

$$J_o = \frac{n[V_o(R_i + R_d) - R_i\Delta\bar{\mu}_H]}{n_H^o F[R_o(R_i + R_d) - R_dR_i]} \tag{3}$$

Here, of course, it is increasing activation of m elements which is taken to lead to dissipation of $\Delta\bar{\mu}_H$. Thus, the ratio of the slopes (ADP-dependent/uncoupler-dependent) for the plots of the dependence of J_o on $\Delta\bar{\mu}_H$ is given as

$$\frac{(\text{Slope})_{\text{ADP}}}{(\text{Slope})_{\text{uncoupler}}} = \left[1 + \frac{mR_d}{nR_e}\right]\left[1 + \frac{R_d(R_i + R_o)}{R_oR_i}\right] \tag{4a}$$

and the intercept ratio as

$$\frac{(\text{Intercept})_{\text{ADP}}}{(\text{Intercept})_{\text{uncoupler}}} = 1 + \frac{R_dR_i}{R_o(R_i + R_d)} \tag{4b}$$

The slope of the J_o vs. $\Delta\bar{\mu}_H$ plot is therefore greater when $\Delta\bar{\mu}_H$ is adjusted with ADP than with uncouplers because in the former case the electrochemical gradient for H^+ in the immediate vicinity of the electron transport chain is dissipated.

These relationships contrast with those for the corresponding model in which coupling is delocalized ($1/R_d \to \infty$, $1/R_i \to 0$), since in the latter case for both ADP and uncoupler treatments,

$$J_o = \frac{nV_o}{n_H^o FR_o}\left[1 - \frac{\Delta\bar{\mu}_H}{V_o}\right] \tag{5}$$

Analogous equations can be devised to construct the dependence of phosphorylation rate, J_P, on $\Delta\bar{\mu}_H$ when $\Delta\bar{\mu}_H$ is adjusted with respiratory inhibitors or uncouplers. Again, the mode of adjustment of $\Delta\bar{\mu}_H$ has a critical bearing on slopes and intercepts of the J_P vs. $\Delta\bar{\mu}_H$ plots. In particular, the slopes are steeper in the presence of inhibitors than in the presence of uncouplers.

The localized coupling model has also been examined with respect to the other potential deviations from delocalized chemiosmotic theory. The apparent rise in n_H^P at low $\Delta\bar{\mu}_H$ is explained again on the basis that diffusive resistance to movement of H^+ from the coupling units to the bulk solution prevents accurate measurement of the electrochemical gradient for H^+ which exists directly across F_oF_1 and is alone responsible for driving ATP synthesis. Thus, according to the electrical equivalent circuit in Figure 1, Equation 1 would be replaced by

$$-\Delta G_P = n_H^P\Delta\bar{\mu}_H\left[1 + \frac{R_d[R_iV_o - \Delta\bar{\mu}_H(R_i + R_o)]}{\Delta\bar{\mu}_H[R_i(R_o + R_d) + R_oR_d]}\right] \tag{6}$$

for the case in which $\Delta\bar{\mu}_H$ is varied with an uncoupler. It will be seen that at high $\Delta\bar{\mu}_H$, Equation 6 simplifies to Equation 1, whereas at low $\Delta\bar{\mu}_H$, the apparent slope ("stoichiometry") approaches ∞.

The results of dual inhibitor titrations are *not* amenable to analysis by the equivalent circuit model unless additional stipulations are made. In particular, it has to be assumed that respiratory inhibitors act to eliminate whole coupling units (F_0F_1 included). Only if this is done is it then possible to model the apparent interaction of inhibitor effects which have given rise to the concept of "cross-talk"[19] between primary and secondary pumps.[17] For example, it has been shown that the near equivalence of rising concentrations of respiratory inhibitors in their effects on $\Delta\bar{\mu}_H$, in state 3 and 4 conditions can be modeled if a steadily increasing fraction (α) of coupling units becomes inoperative. The relevant equations are, for state 3 conditions ($i_p > 0$),

$$\Delta\bar{\mu}_H = \frac{nR_eR_i(1 - \alpha)(R_pV_o + V_pR_o)}{mR_oR_iR_p + \beta[R_iR_p + R_o(R_p + R_i)]} \tag{7}$$

with $\beta = nR_e(1 - \alpha) + mR_d$. For state 4 conditions ($i_p = 0$, R_p high)

$$\Delta\bar{\mu}_H = \frac{nR_eR_iV_o(1 - \alpha)}{mR_oR_i + \beta(R_o + R_i)} \tag{8}$$

Westerhoff et al. show numerically that the efficacy of respiratory inhibitors (as given by $\Delta\bar{\mu}_H = f(\alpha)$, with the absolute magnitudes of the two functions suitably scaled) is similar for state 3 and 4 conditions if $n/m = 10$ and $R_e = R_o = R_d = R_p = 0.1R_i$. Thus, whereas the ratio of state 3 $\Delta\bar{\mu}_H$/state 4 $\Delta\bar{\mu}_H$ is given by Equations 7 and 8 as

$$\frac{[R_pV_o + R_oV_p][mR_oR_i + \beta(R_o + R_i)]}{V_o\,mR_oR_iR_p + \beta[R_iR_p + R_o(R_i + R_p)]} \tag{9}$$

which is dependent on α (subsumed in β) at low R_i and m the ratio simplifies to the α-independent function

$$\frac{R_pV_o + R_oV_p}{R_pV_o} \tag{10}$$

Elimination of coupling units is also proposed as the explanation for observed nonlinear dependence of the phosphorylation rate on $\Delta\bar{\mu}_H$ for cases in which $\Delta\bar{\mu}_H$ is manipulated by addition of respiratory inhibitor. Thus, although the equivalent circuit is a linear one, it is the combination of low $\Delta\bar{\mu}_H$ and coupling unit elimination which can generate nonlinear force-flow relationships.

No attempt has been made throughout the analysis by Westerhoff et al. to fix confidence limits on the parameter estimates used to replicate the general form of experimental observations. Indeed, the analysis restricts itself to a general commentary on the nature of kinetic responses and does not encompass specific data. Nevertheless, the approach does have the virtue of defining some system parameters in mathematical form.

B. Critique of the Localized Coupling Model

In assessing the validity of the mosaic protonic coupling model, it seems appropriate first to focus more critically on the "anomalies" with conventional chemiosmotic theory (discussed earlier) which provoked formulation of the model initially. First, Westerhoff et al.[17]

contend that a major set of observations which is at variance with chemiosmotic theory is the absence of dependence of H$^+$ transport on $\Delta\overline{\mu}_H$ per se. It has, therefore, become common to plot H$^+$ transport (or some related parameter, such as respiratory rate or rate of phosphorylation) as a function of $\Delta\overline{\mu}_H$ and thereby show that transport is not uniquely dependent on $\Delta\overline{\mu}_H$ but on the manner in which $\Delta\overline{\mu}_H$ is adjusted.

This concern over differential effects of modulators of overall driving force arises, at least in part, from the class of model which has formed, explicitly or implicitly, the basis for experimentation. Thus, the dominant concept has been force-flow relationships. The historical basis for this concept appears to have been Mitchell's own extensive use of thermodynamic arguments to advance the chemiosmotic hypothesis. However, it is important to note that the basic tenets of the chemiosmotic hypothesis[4] do not include any statements pertaining to kinetic events and are restricted instead to implications concerning thermodynamic competence. It seems, therefore, unnecessarily restrictive to adopt, at the outset, a model in which a given transport system is constrained to exhibit an identical response to each of the components of $\Delta\overline{\mu}_H$: internal pH (pH$_i$), membrane electrical potential difference ($\Delta\Psi$), and external pH (pH$_o$). In other words, although the energetic relationship between the components of $\Delta\overline{\mu}_H$ can be expressed formally as

$$\Delta\overline{\mu}_H = zF\Delta\Psi + 2.3RT(pH_o - pH_i) \tag{11}$$

and the energetic equivalence of the chemical and electrical components of $\Delta\overline{\mu}_H$ is well established, the precise manner in which these components act upon a transmembrane H$^+$ flux catalyzed by a given transport system cannot be predicted.

Indeed, the futility of using $\Delta\overline{\mu}_H$ as a reference point for description of fluxes has been illustrated for H$^+$-coupled Cl$^-$ transport in *Chara*.[20] Independent manipulation of pH$_i$ and pH$_o$ revealed no unique dependence of Cl$^-$ flux on $\Delta\overline{\mu}_H$, since transport is markedly more sensitive to pH$_i$ than to pH$_o$. It was, however, possible, to account for the differential pH-dependence of transport with a reaction kinetic model of a recycling carrier.

The reaction kinetic approach has also been successful in generating a description of the behavior of primary H$^+$ pumps (Section III), as well as solubilized F$_1$ activity.[21,22] What is clear from all these studies is that it is imperative to initiate modeling with detailed information on the individual reactant concentrations. (For the purposes of modeling electrogenic transport, $\Delta\Psi$ can be considered a reactant.) While the kinetic parameters derived from such modeling must, of course, be bounded by thermodynamic constraints, energetic data alone are not sufficient to enable useful (predictive) kinetic models to be derived.

This point is elementary in the field of steady-state enzyme kinetics: concentrations of all substrates and products of the enzyme should be known for each assay. It is perturbing, then, to note that in many of the research papers from which Westerhoff et al. derive their model, data are presented only in the format: rate vs. driving force. Without information on the separate behavior of the components of $\Delta\overline{\mu}_H$, it is not possible to attempt reaction kinetic modeling. Thus, for critical consideration of the validity of quantitative kinetic models, it should be regarded as of the highest priority that complete information on the behavior of the *components* of $\Delta\overline{\mu}_H$ is readily available.

Classical recycling carrier models can normally be expected to generate nonequivalent kinetics with respect to the various components of $\Delta\overline{\mu}_H$. The current through a 4-state carrier system (Figure 2) can be described[23] as:

$$i_p = zFN \frac{[H^+]_i^z k_{31}^o k_{24} k_{43} k_{12} - [H^+]_o^z k_{42}^o k_{13} k_{34} k_{21}}{[H^+]_i^z [H^+]_o^z k_{31}^o k_{42}^o A + [H^+]_i^z k_{31}^o B + [H^+]_o^z k_{42}^o C + D} \tag{12}$$

out   **in**

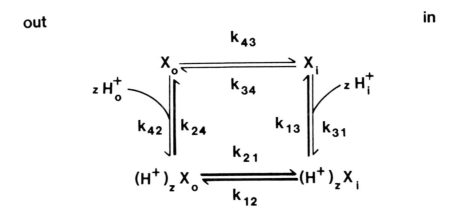

FIGURE 2.   Four-state reaction kinetic model for transport of z protons by a carrier X. Rate constants for transitions between carrier states are indicated as k_{ij}. Note that the charge translocation reaction $((H^+)_zX_i \rightleftharpoons ((H^+)_zX_o)$ and the H^+ binding reactions $(X_i \rightleftharpoons (H^+)_zX_i$ and $X_o \rightleftharpoons (H^+)_zX_o)$ are discrete. The four rate constants marked as heavy arrows are those which have to be large for equivalence of the $\Delta\Psi$ and ΔpH components of the driving force according to the conditions 16.[23]

in which N is the total carrier density, z, the H^+ transport stoichiometry, k_{ij}, identifies the appropriate rate constant in Figure 2,

$$k_{31} = k_{31}^o[H^+]_i^z, \tag{13a}$$

$$k_{42} = k_{42}^o[H^+]_o^z, \tag{13b}$$

(which assumes simple mass action behavior of $[H^+]$ on the transport system), and the coefficients A through D are sums of products of rate constants defined as follows:

$$A = k_{21} + k_{12} \tag{14a}$$

$$B = k_{43}(k_{24} + k_{21}) + k_{12}(k_{43} + k_{24}) \tag{14b}$$

$$C = k_{34}(k_{13} + k_{12}) + k_{21}(k_{34} + k_{13}) \tag{14c}$$

$$D = [k_{43} + k_{34}][k_{24}(k_{13} + k_{12}) + k_{13}k_{21}] \tag{14d}$$

Specific behavior of Equation 12 with respect to $[H^+]_o$ and $[H^+]_i$ will therefore depend on the relative magnitudes of the coefficients A through D.

For an electrogenic primary pump or electrophoretic secondary transport system, the effects of $\Delta\Psi$ can be incorporated into the rate constants k_{12} and k_{21} (which describe the charge-carrying reaction $XH_o^{z+} \rightleftharpoons XH_i^{z+}$) as a symmetric Eyring Barrier:

$$k_{12} = k_{12}^o \exp(zF\Delta\Psi/2RT) \tag{15a}$$

and

$$k_{21} = k_{21}^o \exp(-zF\Delta\Psi/2RT) \tag{15b}$$

Again, it is not difficult to find conditions leading to a unique $\Delta\Psi$-dependent response.[23]

However, conditions can also be found in which flux is *independent* of which particular component of $\Delta\bar{\mu}_H$ is varied. Near equivalence of the effects of $\Delta\Psi$ and pH_o was reported, for example, for F_oF_1-mediated ATP synthesis in *Streptococcus lactis*.[24] The relevant conditions, derived by Hansen et al.[23] and diagrammed explicitly for a 4-state model in Figure 2, take the form of specific size-ordering of the rate constants which comprise the carrier cycle. Specifically, they are as follows:

$$k_{12}^o, k_{21}^o > k_{13}, k_{24} \tag{16a}$$

$$k_{13}, k_{24} >> k_{31}, k_{42} \tag{16b}$$

$$k_{31}, k_{42} > k_{34}, k_{43} \tag{16c}$$

Simply stated, these conditions imply a rather high pK for H^+ binding, and rate-limitation of the carrier cycle primarily in the return reaction of the unloaded carrier $(X_i \rightleftharpoons X_o)$. Applying the conditions 16 a to c, the rather unwieldy Equation 12 describing current through the carrier simplifies to a function depending on $\Delta\bar{\mu}_H$ per se which can be written as

$$i_p = zFN \frac{[H^+]_i^z \exp(zu/2) \, k_{43}Y - [H^+]_o^z \exp(-zu/2) \, k_{34}Z}{[H^+]_i^z \exp(zu/2) \, Y + [H^+]_o^z \exp(-zu/2) \, Z} \tag{17}$$

in which

$$u = F\Delta\Psi/RT, \tag{18}$$

$$Y = k_{24}k_{31}^o k_{12}^o \tag{19a}$$

and

$$Z = k_{13}k_{42}^o k_{21}^o \tag{19b}$$

Equation 17 is seen to be completely equivalent with respect to increasing $[H^+]_o$, decreasing $[H^+]_i$, or corresponding (59 mV/pH unit) shifts in $\Delta\Psi$.

A second important feature to emerge from reaction kinetic models is that there is no basis to expect *linearity* between flux and the components of the driving force. Indeed, there are experimental data indicative of nonlinear dependence of H^+ flux on $\Delta\Psi$ across at least two coupling membranes,[25,26] although background proton conductance appears to be invariant with absolute pH_o in *Paracoccus*[27] and *Streptococcus*.[28] Linear force-flow relationships inevitably result from nonequilibrium thermodynamic and linear equivalent circuit models unless *ad hoc* assumptions are made concerning interaction of separate elements.[17]

Reaction kinetic models for membrane transport exhibit saturability with respect to the components of $\Delta\bar{\mu}_H$. This point can be appreciated quite simply in qualitative terms with reference to H_o^+-binding in Equation 12. At low $[H^+]_o$, carrier turnover is rate-limited by $[H^+]_o$ and inward current displays linear dependence on $[H^+]_o$ (since the second and fourth terms dominate in the denominator). As $[H^+]_o$ is raised and k_{42} increases in direct proportion to $[H^+]_o^z$, carrier recycling becomes rate limited by other reaction constants in the carrier cycle (i.e., those in the second term in the numerator, and those subsumed in A and C). Similar arguments can be advanced for the effects of $[H^+]_i$ and $\Delta\Psi$. Overall, net flux may, though need not necessarily, exhibit michaelian kinetics with respect to the variable parameter, but in any case, the result is saturation of transport as any of the components of $\Delta\bar{\mu}_H$ is raised to high levels.

The inadequacy of the linear approach is amply illustrated by the kinetic behavior, with respect to $\Delta\Psi$, of two well-characterized electrogenic pumps: the plasma membrane H^+ pump of fungi[29] (Section III) and the Na^+ pump of animal cells.[30] In each case, current (net flux) exhibits a quasilinear response to $\Delta\Psi$ close to the equilibrium potential (E_r), but tends to saturation at 100 to 200 mV positive of E_r. The case history of the plasma membrane H^+ pump of characean algae is particularly instructive in the context of the application of linear models. Early electrophysiological characterization of the response of membrane potential to external pH was suggestive of an electrogenic pump functioning around equilibrium. A number of nonequilibrium thermodynamic and equivalent circuit models were advanced which were consistent with a $2H^+$-ATPase.[31-33] Subsequent reaction kinetic analysis[34] has resulted in a convincing explanation of pump kinetics and membrane potential responses to pH_o in terms of $1H^+$-ATPase — a proposal which is strongly supported by the observation that $\Delta\Psi$ can, under some circumstances, exceed the thermodynamic limit bounding the forward operation of a $2H^+$-ATPase.[35]

The demonstrated kinetic versatility of even simple carrier models calls into question not only the significance of the first objection to delocalized chemiosmotic coupling — that there exists no unique correlation between $\Delta\bar{\mu}_H$ and fluxes — but the second and third objections as well: purported evidence for a stoichiometry change from imbalance of ΔG_P and $\Delta\bar{\mu}_H$ at static head, and the conclusion of "cross talk" from dual inhibitor titrations.

The latter point can be dealt with in the context of the arguments outlined above: until there exist precise data on the kinetic response of H^+ fluxes through primary and secondary transport systems to the *components* of $\Delta\bar{\mu}_H$, and corresponding data are presented on the magnitude of these components during dual inhibitor titrations, it will not be necessary to resort to explanations invoking discrete coupling units. For example, if respiratory rate is progressively reduced by titration with an inhibitor of electron transport, there is no basis, as yet, for predicting whether phosphorylation rate will become more or less sensitive to its own inhibitors. This point can be confirmed with further reference to the simple 4-state carrier model (Figure 2 and Equation 12). For the purposes of simplifying the algebra, let us assume $k_{21} \gg k_{12}$. (This restriction only has the capacity to *lower* the versatility of the kinetic response of the model to inhibitors.) Suppose also that a particular inhibitor acts very simply by permitting the reaction k_{43}. (The transition $X_i \rightleftharpoons X_o$ can be viewed as a lumped reaction constant[23] encompassing ADP and P_i binding, ATP release, and carrier recycling.) Then Equation 12 simplifies to

$$i_p = zFN \frac{k_{34}k_{42}k_{13}}{[I] k_{43}^o(k_{31} + k_{13}) + k_{34}(k_{13} + k_{42}) + k_{42}(k_{31} + k_{13})} \tag{20}$$

in which [I] is inhibitor concentration and $k_{43} = [I]k_{43}^o$. We can then write

$$K_i = \frac{k_{34}(k_{13} + k_{42}) + k_{42}(k_{31} + k_{13})}{k_{43}^o(k_{31} + k_{13})} \tag{21}$$

where K_i is the negative of the intercept on the abscissa of a $1/i_p$ vs. [I] plot. The lower the value of K_i, the more sensitive the ATP synthetase is to inhibition by I. As k_{42} ($= [H^+]_o^z k_{42}^o$) is lowered there will either be little effect on K_i (if $k_{34}k_{13}$ is high) or K_i will decrease, since k_{42} appears only in the numerator of Equation 21. Similarly, if k_{31} ($= [H^+]_i^z k_{31}^o$) is raised, K_i will either be reduced, or there will be little effect. Either of these manipulations corresponds to lowering of $\Delta\bar{\mu}_H$. Thus, there appears to be no basis for the conclusion[18] that "on a strict chemiosmotic basis . . . the rate of ATP synthesis should become less sensitive to the initial titres of an inhibitor of the ATP synthase as the rate of electron transport is progressively reduced by titration with an inhibitor or by restriction of substrate supply''.

Observations suggesting that n_H^P is $\Delta\bar{\mu}_H$-dependent have been questioned recently in a rigorous study.[36] These and other experimental arguments[18] notwithstanding, it is important to note that the reversal potential of a carrier-type transport system is not necessarily a crisply defined parameter from an experimental point of view. Thus, it has been shown[23] that there can exist large voltage spans, which encompass E_r, in which almost no current flows through the transport system. Such spans might, conceivably, lead to erroneous estimates of E_r if this parameter is simply derived by extrapolation, rather than reversal of the system.[37] Similar features emerge from analysis of the control of current flow by ligand concentration.[38] The extent to which functional reversibility appears clearly around E_r is dependent on the transition rates of the various carrier states; some of these, in turn, are dependent (via mass action) on the prevailing ligand concentrations. Thus, although F_0F_1 is functionally reversible, it is essential to demonstrate such reversibility in the same conditions as those of static head experiments in order to eliminate the possibility that the transport system is at kinetic steady state of near-zero current flow.

A central question, then, is whether H^+ transport through coupled F_0F_1 can be considered compatible with conventional carrier models. Incorporation of purified F_0 in planar lipid membranes has indicated some channel characteristics: rapid turnover (10^7/sec) and a reasonably linear current-voltage curve.[39] However, these experiments were performed at pH 2.2 (for technical reasons) and flux was not rate limited by the blocking action of F_1 as it is in native membranes. On the other hand, F_1 itself certainly exhibits classical enzyme kinetics with respect to ATP hydrolysis, and it is in the context of the role of H^+ in driving conformational changes in F_1 that reaction kinetic analysis might most usefully be applied. Perhaps, therefore, in contrast to the simple 4-state model on which attention was focused earlier, modeling should more realistically be initiated on the basis of a H^+ channel with multiple conformational states.[40]

An elegant treatment of such models, which form a hybrid between traditional conceptions of channel and carrier models, is given by Läuger,[8] who considers the case of incomplete coupling between H^+ flows and ATP synthesis. This leads automatically to a distinction between the coupling ratio (defined as the number of H^+ transported per ATP synthesized) and the stoichiometric ratio (the number of binding sites directly involved in transport). Whereas the latter parameter is mechanism-dependent and constant, the former might be expected to vary with $\Delta\bar{\mu}_H$. Thus, as Läuger[8] points out, the reversal potentials for ATP synthesis and H^+ transport might not be identical. This model also neatly explains the kinetic evidence[24] which might be taken to suggest that H^+ passes through the voltage drop before interacting with the ATP synthase.[37]

C. Linear or Nonlinear Models to Describe H^+ Fluxes in Energy-Coupling Membranes?

In conclusion, then, I believe there should be more detailed consideration of what constitutes a "minimal model" for H^+ fluxes across energy coupling membranes. Nonequilibrium thermodynamic models have the virtue of generating simple linear equations. Such models are, however, phenomenological and are difficult to place in the physical context of membrane protein structure. Furthermore, the simple linear equations are not capable of accounting for the kinetics of H^+ transport unless a complex physical and functional interaction of primary and secondary H^+ transport systems is envisaged. Thus, analysis which confines itself at the outset to the postulate of linearity and equivalence of the components of $\Delta\bar{\mu}_H$ inevitably generates anomalies between what is viewed as the predictions of chemiosmotic theory on one hand, and real data on the other.

Reaction kinetic models, in contrast, clearly result initially in more complex equations. However, they are easy to relate to general principles of enzyme action and to the kinetic behavior of other transport systems (Sections III and IV). The rate equations are flexible enough to describe nonlinearity between force and flux, and account for the kinetic non-interchangeability of components of the driving force. Potentially such models are testable, especially if nonsteady-state kinetics can be applied.[12] Finally, Chapman et al.[41] have stated: "The speed and availability of digital computers now permit almost unrestricted quantitative testing of kinetic schemes and interpretative ideas concerning biological transport processes . . . To this extent there is no longer any need to persist with simplifying mathematical devices (e.g., idealized Michaelis-Menten schemes, linear non-equilibrium thermodynamics . . .) to analyze complex transport systems."

III. THE EUKARYOTIC PLASMA MEMBRANE H^+ ATPase

A. Empirical Derivation of Pump I-V Curve

With the possible exception of some marine-dwelling organisms and other halophytes, an electrogenic H^+ pump is widespread in the plasma membrane of eukaryotic microorganisms.[42] The protein is an ATPase, M_r 95,000 to 105,000, and bears many similarities in its in vitro characteristics to the K^+ ATPase of *Escherichia coli*, Na^+/K^+ATPase of animal cell plasma membranes, the Ca^{2+} ATPase of sarcoplasmic reticulum, and the H^+/K^+ ATPase of gastric mucosa.[7] All, for example, are extremely sensitive ($K_i < 10\ \mu M$) to the phosphate analog, orthovanadate. These enzymes are known as E_1E_2 ATPases, a terminology which is intended to reflect their capacity to exist in two well-defined conformational states. Kinetic models for the activity of the solubilized H^+-ATPase have been derived for the yeast *Saccharomyces cerevisiae*[43] and for the mycelial fungus, *Neurospora crassa*.[44]

A feature common to all E_1E_2 ATPases from microorganisms is their capacity to pump H^+ electrogenically. Thus, *in situ*, pumping of H^+ occurs from the inside of the cell to the outside, and no other ion appears to be involved.[7] This fact introduces an additional dimension to kinetic modeling of H^+ transport, as it does for treatment of F_0F_1 ATPase, because, besides ligand concentrations, the effects of $\Delta\Psi$ must be incorporated into any model. The effects of $\Delta\Psi$ on pump current are described by a current-voltage (I-V) curve for the pump.

Since the pump passes current, its activity can, in principle, be monitored with electrophysiological techniques. Impalement of most microorganisms with microelectrodes is difficult or impossible. Fortunately, hyphae of *Neurospora* are readily amenable to impalement, and this has enabled Slayman and colleagues to undertake considerable modeling of the electrical kinetics of H^+ pumping.

Early work[45] attempting to derive an I-V curve for the H^+ pump in *Neurospora* relied on a subtraction method. A control I-V curve was obtained for the plasma membrane with a fully functional H^+ pump. CN^- was then added to inhibit the pump via ATP depletion, and a second I-V scan run. The difference between the two curves (fit by polynomials) was taken to represent the I-V curve of the pump. The averaged difference curves were finally fitted to a single exponential function, revealing that pump current increased with increasingly positive membrane voltage, as might be expected of a transport system responsible for an outward current.

Two features of this method were rather unsatisfactory. First, the difference curve method assumes that all pumps are inhibited during CN^- treatment. As was recognized at the time,[45] residual ATP probably permitted the pump to operate at 25% of control rate. However, as has been pointed out quite forcibly,[41] the consequences of incomplete and reversible inhibition can be more serious than the introduction of a scaling error in the pump I-V relation. Operation of the pump at a reduced rate but with changed adenine nucleotide concentrations can also result in a *shape* change of the I-V curve, and even an incorrect estimate of E_r. The second unsatisfactory feature of the difference curve method was that polynomial or exponential fits, being essentially phenomenological, could not be deemed to have extracted all available kinetic information from the I-V data.

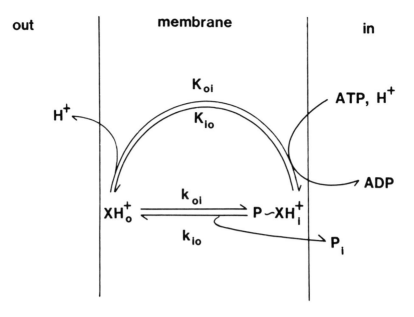

FIGURE 3. Two-state representation of an electrogenic proton pump of the E_1E_2 type. Charge translocation reactions are designated by their rate constants k_{io} and k_{oi}. All voltage-insensitive reactions (ligand binding, return of uncharged carrier) are lumped into the gross reaction constants K_{io} and K_{oi}.

B. Utility of Two-State Reaction Kinetic Model

Application of reaction kinetic models to whole membrane I-V curves overcomes both these difficulties. Hansen et al.[23] have shown that any multistate model for an electrogenic transport system in which transmembrane charge translocation occurs in just a single step (class-I models) can, for the purposes of description of any I-V curve, be reduced to a simple 2-state model. Thus, all the voltage-independent reactions (comprising ligand binding, return of the uncharged carrier, etc.) can be subsumed into two gross reactions (Figure 3). Current carried by such systems can then be written

$$i_p = zF(k_{io}N_i - k_{oi} N_o) \tag{22}$$

in which the concentrations of the two carrier states, $P \sim XH_i^+$ and XH_o^+, are designated N_i and N_o, respectively. For the steady state ($dN_i/dt = 0$, $dN_o/dt = 0$),

$$0 = N_i(k_{io} + K_{io}) - N_o(k_{oi} + K_{oi}) \tag{23}$$

and, for the model in Figure 3, as drawn,

$$N = N_i + N_o \tag{24}$$

Substituting Equations 23 and 24 into Equation 22,

$$i_p = zFN \frac{K_{oi}k_{io} - K_{io}k_{oi}}{K_{io} + K_{oi} + k_{io} + k_{oi}} \tag{25}$$

The effect of $\Delta\Psi$ can be introduced into Equation 25 via Equations 15.

The 2-state model defines a wide variety of shapes of I-V curves. The ratio ($\sqrt{k_{oi}k_{io}}$)/

$(K_{oi} + K_{io})$ is particularly crucial to shape. If this ratio is large, the I-V curve can be described as a hyperbolic tangent, i.e., with a large slope conductance around the equilibrium potential. If, alternatively, the ratio is small (carrier reloading rate limiting) inflections in the I-V curve can occur between the two regions of saturating current, even to the extent of generating a large span of voltage-insensitivity around E_r.

The 2-state model embodies six variables $(z, N, k_{io}, k_{oi}, K_{io}, K_{oi})$. Description of the whole membrane I-V curve also has to take into account a pathway for current return to the cell, and, as a first approximation, an ohmic "leak" has been assumed. Thus, with two additional parameters needed to describe the leak (its equilibrium potential and conductance), the minimum number of variables needed to describe a membrane I-V curve by this formalism is 8. This number can be reduced to 6: experience with modeling the *Neurospora* proton pump has demonstrated that in this case z should be fixed at $+1$, and N, since it is only a scaling factor, is maintained constant at 10^{-12} mol/cm^2. Nevertheless, with this large number of degrees of freedom, it would obviously be unwise to place any credence in the validity of the kinetic estimates which might emerge from the fitting of a single I-V curve.

However, the situation can be improved considerably by *joint* fitting. Thus, the strategy is to consider the control and treatment I-V curves together, and to fit the data allowing a change in only one parameter to describe the differences between the curves. In practice, then, each parameter in turn is allowed to vary in response to the treatment; all other parameters are constrained to maintain identical values for the two curves (though these values are nevertheless optimized by the fitting routine).

Reanalysis of the original control and $(+CN^-)$ I-V curves[29] revealed that the effects of CN^- could satisfactorily be described as occurring via a change in one gross reaction constant: K_{oi}. None of the other reaction constants nor the leak parameters, when varied alone yielded a good fit. The modeling resulted in the following estimates for the pump reaction constants: $k_{io}^o = 2.8 \times 10^{-6}$/sec, $k_{oi}^o = 1.8 \times 10^{-1}$/sec, $K_{oi} = 3.5 \times 10^2$/sec, $K_{io} = 2.4 \times 10^4$/sec. In the presence of CN^-, K_{oi} falls to 1.3×10^2/sec. This CN^--induced change in K_{oi} accords nicely with naive prediction, since it is K_{oi} which presumably subsumes ATP binding.

Another interesting feature of the results is as follows. E_r can be described directly from Equation 25 for $i_p = 0$ as[23]

$$E_r = \frac{RT}{zF} \ln \frac{k_{oi}^o K_{io}}{k_{io}^o K_{oi}} \tag{26}$$

For a 2.7-fold drop in K_{oi} in the presence of CN^-, reversal potential changes only by 8% (from -308 to -283 mV). However, the effect on (positive) saturation current i_{sat^+} of the pump is a 2.7-fold drop, since[23]

$$i_{sat^+} = NK_{oi} \tag{27}$$

Once again (see Section II), the discrepancy between the thermodynamic and kinetic approaches to the description of H^+ transport is therefore highlighted: if generated by change in the appropriate reaction constants, only small shifts in E_r are capable of causing quite large changes in pump current over voltage spans well removed from E_r.

Orthovanadate is rather similar to CN^- in its effects on pump kinetics.[46,47] Again, K_{oi} was the reaction constant which, when free to vary, best described the kinetics of inhibition. A primary effect of vanadate on K_{oi} might be anticipated if vanadate interferes with ATP hydrolysis. The reduction in i_{sat^+} is also 2.5- to 3-fold, with only a small shift in E_r.

C. Reserve Factors and Their Effect on Model Parameters

The kinetics become somewhat more complicated, however, for the effects of pH_i and

pH_o on the pump I-V curve. Recall that although the 2-state model is capable of describing any I-V curve for a class-I system, the model nevertheless subsumes a large number of carrier states. The corollary of this statement is that for any given I-V curve, these subsumed states remain hidden and experimentally inaccessible. The consequence of this lumping is that the law of conservation of mass, which is the basis for Equation 24, is violated. Formally, this situation can be remedied by the introduction of so-called reserve factors[23] which recognize the potential existence of hidden states and are themselves combinations of the voltage-independent rate constants such that

$$N = r_i N_i + r_o N_o \tag{28}$$

For a "real" 3-state model, in which state N_3 is linked to state $N_1 (= N_i)$ and $N_2 (= N_o)$ by the voltage-insensitive reaction constants k_{13}, k_{31}, and k_{23}, k_{32}, respectively,

$$r_i = 1 + \frac{k_{13}}{k_{31} + k_{32}} \tag{29a}$$

and

$$r_o = 1 + \frac{k_{23}}{k_{31} + k_{32}} \tag{29b}$$

The more accurate form of the reduced 2-state equation from a multi-state model then becomes, from Equations 22, 25, and 28

$$i_p = zFN \frac{k_{12}K_{21} - k_{21}K_{12}}{r_i(k_{21} + K_{21}) + r_o(k_{12} + K_{21})} \tag{30}$$

Note that this equation differs from Equation 8 of Hansen et al.[23] in eliminating reserve factors from the definition of N. Comparison of Equation 30 with Equation 25 yields the following definitions:

$$k_{io} = \frac{k_{12}}{r_i}, \quad k_{oi} = \frac{k_{21}}{r_o}, \quad K_{io} = \frac{K_{12}}{r_i}, \quad K_{oi} = \frac{K_{21}}{r_o} \tag{31a-d}$$

Hansen et al. also derived expressions for K_{12} and K_{21} in terms of a 3-state model. These are given as

$$K_{12} = \frac{k_{13}k_{32}}{k_{31} + k_{32}} \tag{32a}$$

and

$$K_{21} = \frac{k_{31}k_{23}}{k_{31} + k_{32}} \tag{32b}$$

This then enables definitions of the output parameters of a 2-state fit (k_{io}, k_{oi}, K_{io}, K_{oi}) to be expressed in terms of a real 3-state model by substitution of Equations 29 and 32 into Equations 31:

$$k_{io} = \frac{k_{12}(k_{31} + k_{32})}{k_{31} + k_{32} + k_{13}} \tag{33a}$$

$$k_{oi} = \frac{k_{21}(k_{31} + k_{32})}{k_{31} + k_{32} + k_{23}} \tag{33b}$$

$$K_{io} = \frac{k_{13}k_{32}}{k_{31} + k_{32} + k_{13}} \tag{33c}$$

$$K_{oi} = \frac{k_{31}k_{23}}{k_{31} + k_{32} + k_{23}} \tag{33d}$$

The most important point, then, is that in fitting the 2-state model to I-V data, we might expect the effect of manipulation of a given voltage-insensitive reaction to manifest itself in some gross reaction constants not directly subsuming the manipulated reaction. For example, for a "real" 3-state model in which

$$k_{13}, \quad k_{23} >> (k_{31} + k_{32}) \tag{34}$$

the effect of a rise in k_{32} will appear not only as an increase in K_{io} (into which k_{32} is directly subsumed) but also as a possible increase in k_{oi} and k_{io}. Similarly, a rise in k_{31} has the capacity to elicit increases in k_{oi} and k_{io} as well as in K_{oi}.

It became apparent during analysis of experiments in which either pH_i or pH_o was changed, that satisfactory fits could not be obtained with a 2-state model in which only one pump parameter was free to vary with pH. Thus, in the case of a reduction in pH_i (from a control value of 7.12 to 6.65) it was necessary to let not only K_{oi} vary (as expected for mass action binding of H_i^+ to the pump), but also k_{io}^o.[48] This effect contrasts with the results of CN^- and vanadate treatment, but can be rationalized by taking into account model reduction and the nature of the reserve factors. (In addition, a large increase in leak conductance was noted; that should serve to modulate pump velocity via voltage effects.) The overall increase in K_{oi} is 2.5-fold, and for k_{io}^o it is 1.1-fold. Thus, as with the CN^- results, a large change in i_{sat^+} (Equation 27) is apparent.

Kinetic models must always be bounded by thermodynamic constraints. With respect to the effects of $[H^+]_i$, it is gratifying to note that despite the large shift in i_{sat^+}, the overall change in E_r was only 24 mV (see Equation 26), which is in close accord with the theoretical value for a 0.43 unit pH change (25 mV). The time course for the shift in E_r was shown to agree well with the time course for change in pH_i.

In an exactly analogous manner, the effects of decreased pH_o show up as an increase in both K_{io} and k_{oi}^o.[49] The naive interpretation from the 2-state model would, of course, anticipate effects of $[H^+]_o$ to be seen only in K_{io}, which subsumes H_o^+ binding (Figure 3). K_{oi}, and therefore i_{sat^+}, remains unaffected by $[H^+]_o$. Nevertheless, pump current over the physiologically significant range of membrane potential (-150 to -230 mV) decreases significantly during lowering of pH_o from 5.8 to 3.7. Consistent with thermodynamic predictions, this pH change induces a positive shift in E_r of 123 mV. Similar effects of pH_o on the plasma membrane H^+ pump in characean algae have also been reported as a result of the application of reaction kinetic models.[34,50]

D. The Need for Higher-State Models

It is apparent that a major drawback of fitting the effects of pH_i and pH_o is the absence of kinetic predictability. Because H^+ binding manifests itself over at least two gross reaction constants and the values of the reserve factors are not immediately known, it is not possible, for example, to anticipate the effect of a shift in pH in the opposite (alkaline) direction because the reserve factors will change in magnitude. A simple way around this dilemma is to consider fitting higher-state models. The minimum number of states to describe the effect of a ligand in terms of a single rate constant varies with the relative magnitude of the partial reaction rates.[23] We have already seen that for CN^- and vanadate, a 2-state model suffices. As many as four states may be required if the ligand binding reaction is fast and separated from a fast voltage-sensitive reaction by two slow reactions. The order of model to be used must be determined empirically by progressive elimination of lower-order models, until the effects of a ligand can be described by variation in just one reaction rate.

Taken individually, pH_i and pH_o effects can satisfactorily be localized by fitting with a 3-state model.[51] In the case of lowered pH_i, the reaction constant generating the best fit was, equally with one other reaction constant, k_{31}. This has enabled predictions to be made concerning the effect of raised pH_i on pump current. As might be expected from the foregoing discussion, the major prediction is a dramatic fall in i_{sat^+}, and this has indeed been confirmed by experiment.[51]

For lowered pH_o, all reaction constants in the 3-state model can be constrained except k_{32}, which changes in approximate proportion to $[H^+]_o$. Here, however, the rather surprising prediction is that lowering of $[H^+]_o$ should have no effect on pump current in the measurable range (positive of -300 mV). Experiments in which pH_o was raised from pH 5.8 to 8.2 confirmed this prediction: there is no effect on the membrane I-V curve.[51]

Reconciling the different effects of pH_i and pH_o into a single model, with only one reaction constant free to change for pH_i and one for pH_o, necessitates the use of four carrier states, such as is diagrammed in Figure 2. This enables the effect of $[H^+]_i$ to be localized by variation of k_{31}, and $[H^+]_o$ by k_{42} for joint fitting of experiments in which pH_i or pH_o is changed.[51] The kinetic constants which emerge provide a conceptual basis for interpreting effects of pH on pump current.

First, the ratios k_{13}/k_{31}^o and k_{24}/k_{42}^o can be used to estimate pKs for the internal and external H^+ binding sites. These pKs turn out to be 5.4 for the internal site and 2.9 for the external site. There is significance in these values for the role of the pump in cytoplasmic pH homeostasis. The normal pH_i is 7.2. Thus, the H_i^+ binding site exists predominantly in the dissociated form and availability of loaded carrier to cross the membrane will effectively rate limit proton pumping. Pump current will therefore tend to be sensitive to changes in pH_i whether pH_i is raised or lowered from its control value. The pK of 2.9 for the external site also implies that the site exists predominantly in the dissociated form at all reasonable pH_o, as well. Lowering pH_o from the control value of 5.8 can have an effect on pump current as the significance of the back reaction $[H^+]_o k_{42}^o$ is increased. Raising pH_o has little or no effect because dissociation is already almost complete at pH 5.8. These arguments can be quantitatively confirmed by reference to Equation 12, remembering that i_p is positive for this transport system (i.e., only the first term in the numerator is relevant).

A second feature emerging from the 4-state model is that the nonlinear and nonequivalent effects of the components of $\Delta\bar{\mu}_H$ on the pump current can be accounted for.[51] Calculations reveal that increasing $\Delta\bar{\mu}_H$ by 58 mV from control conditions results in almost zero current flow if the pH_i or $\Delta\Psi$ components are changed, but little effect if the pH_o component is varied. Conversely, decreasing $\Delta\bar{\mu}_H$ by 58 mV results in 2.3-fold stimulation of pump current if $\Delta\Psi$ is manipulated, a 10-fold change for pH_i, and no change for pH_o.

The kinetic constants derived from the 4-state model also allow cross-referencing between the electrophysiological and biochemical approaches; the pH profile for ATPase activity of the solubilized enzyme can be predicted.[52] The simple conditions are $\Delta\Psi = 0$ and $pH_i = pH_o$. The projected pH profile increases monotonically with decreasing pH, showing saturation between pH 4.5 and 5. However, the measured curve shows a distinct pH optimum around pH 6.5.[53] One reason for the discrepancy might be acid-inactivation of the isolated enzyme, and it is interesting to note that correction for acid-inactivation of *Saccharomyces* ATPase[54] results in tolerably close agreement to the pH profile predicted by the reaction kinetic model.

E. Future Developments

It will be apparent that two assumptions underlying this reaction kinetic modeling of plasma membrane H^+ transport in *Neurospora* are (1) that the current-carrying nonpump pathways behave ohmically and (2) that the energy barrier for charge translocation is symmetric across the membrane. So far, modeling has generated self-consistent results, despite these unjustified assumptions. Thus, application of more refined models[8] for this particular electrogenic pump might be premature. Further analysis of pump kinetics may have to await the development of a specific and rapidly acting inhibitor analogous to ouabain, as well as a greater characterization of the leak conductance. Alternatively, reconstitution of the purified H^+-ATPase into black lipid membranes or giant liposomes might prove an easier way of separating the pump current from that of the other electrically active pathways in the native membrane.

IV. H^+-COUPLED TRANSPORT

A. Influence of Techniques on Conception of Models

As the principal method of $\Delta\bar{\mu}_H$ dissipation at the plasma membrane of eukaryotic organisms, and certainly an important pathway in prokaryotes, H^+-gradient coupled transport systems deserve consideration as part of the chemiosmotic H^+ circuit. Contrary to the situation pertaining to the primary $\Delta\bar{\mu}_H$-generating and -consuming systems in energy-coupling membranes, there has been no debate concerning the applicability of conventional enzymic schemes to gradient-coupled transport. This, almost certainly, has arisen from the nature of the techniques used to study these respective systems. For H^+-translocating electron transport chains and for ATPases in which H^+ is the sole translocated ligand, investigators must rely on estimates of net flux transmitted by the system — whether this is measured as a change in bulk phase activity, as an electrical current, or even as phosphorylation rate. Gradient-coupled systems, on the other hand, are normally amenable to radioisotopic analysis of the flux of at least one of the transported ligands, and thus unidirectional fluxes can be measured as well. Since these unidirectional fluxes are always saturable, and most frequently exhibit michaelian kinetics, the applicability of some sort of enzymic formalism has been obvious. The consequent nonlinearity of the flux/driving force relationship has not presented itself as a conceptual problem to investigators, and despite the fact that many of the systems are reversible, the temptation to provide a nonequilibrium thermodynamic interpretation of kinetics has usually been avoided.

Furthermore, since gradient-coupled transport was first postulated for epithelia,[55] in which Na^+ rather than H^+ is the "driver ion", many of the concepts in modeling owe more to developments in the field of Na^+-coupled transport (and in particular to the classic review of Schulz and Curran[56]) than to explicit recognition of the place of H^+-coupled transport in chemiosmotic H^+ circuits. Again, availability of $^{22}Na^+$ as a probe for unidirectional Na^+ transport has meant that Na^+-coupled systems can be studied in greater depth than their H^+-coupled counterparts. However, so far as can be ascertained, there are no fundamental differences in the general kinetic characteristics of Na^+-coupled systems in animals and bacteria and of H^+-coupled systems, thereby suggesting that similar classes of models should be applicable.

B. Rapid Equilibrium Binding of Ligands to a Carrier

Early formalism assumed ordered binding carriers of the type shown in Figure 4A. It was recognized that binding order of the two ligands can play a crucial part in the determination of kinetics, and indeed, Schultz and Curran[56] suggested that the effects of driver ion concentration on solute influx might be used to deduce binding order.

In a further development of reaction kinetic modeling for coupled transport, Heinz and

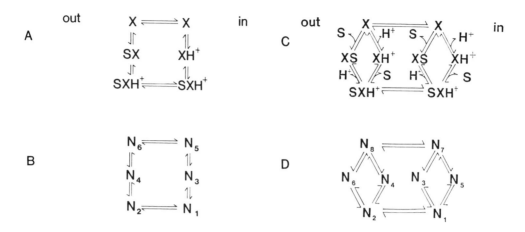

FIGURE 4.   Reaction kinetic models for symport of a solute (S) with protons. (A) Ordered binding model for S-H$^+$ symport on carrier, X. With respect to entry of S, the model is First-on, First-off (FF), and charge translocation is depicted as occurring on the loaded form of X. Three other possible binding orders exist: FL, LF, and LL. (B) Generalized representation of A, with densities of carrier states denoted N_j. Rate constant for the reaction step from N_i to N_j can be written k_{ij} (see text Equations 35 to 40). (C) Random binding model for S-H$^+$ symport on carrier, X. This representation can be viewed as a general scheme, of which the model in A and its three congeners are particular examples for specific size-ordering of rate constants in the random binding model. (D) Generalized representation of C.

co-workers[57,58] separately considered models in which components of the electrochemical gradient of the ion driving solute uptake might either selectively affect the affinity of the carrier for the solute, or the intrinsic carrier mobility. Specific application of these so-called affinity and velocity models to transport kinetic data from microorganisms has been considered previously.[59]

With only a few exceptions, all detailed modeling of gradient-coupled transport was constrained at the outset to rapid equilibrium binding of ligands to the carrier. In other words, it was assumed that translocation steps ($N_1 \rightleftharpoons N_2$ and $N_5 \rightleftharpoons N_6$ in Figure 4B) were all small in comparison with ligand-binding steps. This constraint appears to have had several origins: that the carrier molecules themselves might diffuse rather slowly across membrane lipid (we now know that transport is likely to be catalyzed by subtle intramolecular rearrangements of a carrier molecule, which could be rapid); that for most enzymes, the rapid equilibrium assumption provides a satisfactory basis for description of catalysis; and that rate equations were cumbersome and needed simplifying in order to appreciate their salient features. Generally, in the late 1970s, most descriptions of kinetic models for gradient-coupled transport recognized this assumption, but merely commented that rapid equilibrium binding is "the conventional assumption".

C. Nonrapid Equilibrium Models

It is now clear that the *a priori* assumption of rate limitation of transport by transmembrane reaction steps is not justified. In equilibrium exchange experiments, Hopfer and Groseclose[60] demonstrated that Na$^+$ dissociation from the Na$^+$-glucose symporter of kidney brush border vesicles occurred at only 10 to 20% of the transmembrane reaction rate. Subsequently, more detailed analysis of data from the same experimental system[61] has amply confirmed the conclusion that the rapid equilibrium approach should be rejected in this case. Thus far, there are no compelling arguments that the rapid equilibrium assumption should be retained for H$^+$-coupled systems, and in at least one well-documented case in microorganisms (that of H$^+$-coupled SO$_4^{2-}$ transport in *Penicillium*) the conclusions concerning ligand binding order[62,63] rest wholly upon the (unjustified) rapid equilibrium assumption.[38] Until further evidence concerning rate-limitation by transmembrane steps is forthcoming, it is suggested that the assumption be abandoned for future modeling work.[61]

The complete rate equation has been derived[38] describing transport of isotopic substrate (*S) through the reaction scheme in Figure 4A, as well as the equations for the seven congeners of this model: the three alternative binding orders, and the class of four models in which charge translocation was taken to occur on the unloaded form of the carrier. With no *S on the *trans* side of the membrane (the normal experimental condition: presence of nonisotopic S on the *trans* side is not excluded) such reaction schemes always generate rate equations of michaelian form. Expressed in terms of the component rate constants, the relationships for J_{max} and K_m are indeed cumbersome. K_m, for example, contains 21 terms in the numerator, and 15 terms in the denominator, each of which is the product of 5 different rate constants. These expressions are, however, greatly simplified if one of three experimentally attainable conditions is imposed: saturating $[H^+]_o$, very low $[H^+]_i$, or saturating negative $\Delta\Psi$. The resulting relationships for J_{max} and K_m are simple enough to enable evaluation of the effect on them of the other two components of $\Delta\bar{\mu}_H$. Key reaction constants, whose relative magnitude determines the specific response, can then be identified. For example, for the model in Figure 4A, after the simplifying conditions of saturating (negative) $\Delta\Psi$ are imposed, the response of the two michaelian parameters to $[H^+]_o$ can be written as

$$J_{max} = \frac{[H^+]_o\, k_{42}^o k_{13} k_{35} k_{56}}{[H^+]_o\, k_{42}^o[(k_{31} + k_{13})(k_{53} + k_{56}) + k_{35}(k_{13} + k_{56})] + k_{13}k_{35}k_{56}} \tag{35}$$

and

$$K_m = \frac{([H^+]_o\, k_{42}^o + k_{46})[k_{53}k_{65}(k_{13} + k_{31}) + k_{13}k_{35}(k_{56} + k_{65})]}{k_{64}^o\{[H^+]_o\, k_{42}^o[(k_{31} + k_{13})(k_{53} + k_{56}) + k_{35}(k_{13} + k_{56})] + k_{13}k_{35}k_{56}\}} \tag{36}$$

Similar relations hold for the other two simplifying conditions and the other three possible binding orders. Inspection of Equations 35 and 36 reveals that raising $[H^+]_o$ can have a number of highly specific kinetic effects, depending on the size ordering of the component rate constants. For an effect on J_{max} alone,

$$k_{13}k_{35}k_{56}, \quad k_{46} > k_{42}^o[H^+]_o \tag{37}$$

and on K_m alone,

$$k_{46} > k_{42}^o[H^+]_o > k_{13}k_{35}k_{56} \tag{38}$$

Joint effects can appear on J_{max} and K_m, which change in parallel with each other if

$$k_{13}k_{35}k_{56} > k_{42}^o[H^+]_o > k_{46} \tag{39}$$

or in opposing directions if

$$k_{46} > k_{42}^o[H^+]_o \simeq k_{13}k_{35}k_{56} \tag{40}$$

This kinetic versatility is retained for most simplifying conditions in all the other binding schemes. Thus, the varied effects of driver ion concentration on the kinetics of gradient-coupled transport in microorganisms[63-66] might easily be explained on the basis of simple reaction kinetic models exhibiting ordered binding. Analogously, kinetically varied effects of $\Delta\Psi$ and $[H^+]_i$ can be simply encompassed by ordered binding models.

The kinetic versatility of these models is potentially an impediment in the determination of binding order or site of charge translocation. However, it has been shown[20,38] that by application of specially chosen experimental conditions (e.g., $[H^+]_o$ and $\Delta\Psi$ saturating, $[S]_i = 0$), the rate equations become model-specific. For example, for these particular conditions, the kinetic response to variation of $[H^+]_i$ will allow progressive elimination of at least some of the models.

A further interesting ramification of even these simple ordered binding models is their capacity to describe kinetic properties of transport for which analysis of rapid equilibrium binding models had seemed to require "special mechanisms". One of these special mechanisms is the proton well, which is supposed to explain how an electrical potential can be transduced into a proton chemical gradient. Taking data on H^+-sugar symport in the microalga *Chlorella*,[67,68] it was shown that apparent $\Delta\Psi$-induced shifts in pK for H_o^+ binding could in fact be replicated by an ordered binding model in which $[H^+]_o$ and $\Delta\Psi$ act on *discrete* reaction steps. The confusion arises because the $K_{1/2}$ of transport for H_o^+ is only indirectly related to the pK of the H_o^+ binding site.

Ordered binding reaction kinetic models can also explain some,[69] but probably not all[70] data which had been cited in favor of a pH_o-dependent change in H^+/solute stoichiometry. The observations center around (1) the capacity of $\Delta\Psi$ to "drive" transport at some pH_o, but not at others, and (2) the capacity of increased pH_i to drive transport at pH_o where $\Delta\Psi$ is not effective. Rationalization of these results with a mechanism involving constant stoichiometry can be achieved if it is recognized that the rate equations are saturable functions of the components of $\Delta\bar{\mu}_H$. Thus, the particular component of $\Delta\bar{\mu}_H$ which rate limits transport can itself be determined by pH_o: at some values of pH_o the crucial rate limiting factor is $\Delta\Psi$, while at other pH_os, it is pH_i.

There have been comparatively few studies of the I-V relations of gradient-coupled transport systems.[71,72] (A preliminary theoretical treatment based on nonrapid equilibrium ordered binding models has been given.[73]) However, it is a common observation that many gradient-coupled transport systems in microorganisms appear to cease net transport at driving forces well in excess of equilibrium, especially at high $[S]_o$.[74] This phenomenon can be understood in terms of effective kinetic limitation of the carrier by $[S]_i$ (transinhibition). A specific theoretical example has been presented[38] of a reaction kinetic model in which significant net flux occurs only when the driving force is in excess of 150 mV if S accumulates on the *trans* side. Transinhibition might eliminate the need to resort to solute leaks internal to the carrier ("slip") as an explanation for the failure of $n\Delta\bar{\mu}_H$ and $\Delta\bar{\mu}_s$ to come to equilibrium at high $[S]_i$.

D. The Lactose-H^+ Carrier of *Escherichia coli*

Of all the gradient-coupled transport systems in microorganisms, that which has been subjected to the most detailed investigation is the H^+-lactose carrier of *Escherichia coli*. The H^+/lactose stoichiometry is 1.[75] In general, the kinetics of the system are adequately described by rapid equilibrium carrier models with $\Delta\Psi$ acting on a discrete transmembrane reaction and $[H^+]_o$ via mass action on H^+ binding.[66,76] This type of model applies particularly for transport of the lactose analog β-D-galactosyl-1-thio-β-D-galactoside (GalSGal). Since exchange is unaffected by imposed $\Delta\Psi$, and yet influx is enhanced in these conditions, it is concluded that transmembrane transit of the unloaded carrier is voltage-sensitive, i.e., that it bears a negative charge.[76] pH effects on kinetics can be explained entirely by mass action effects on a 4-state carrier (fully loaded or unloaded). As in the case of the 6-state model analysis (above), it is not necessary to resort to slip to explain the observation that the accumulation ratio for GalSGal decreases as $[GalSGal]_o$ is increased. However, in the case of the lactose carrier, it is supposed that at high $[GalSGal]_o$, the loaded carrier cycles back and forth on either side of the membrane without unloading.[77]

Structural work on the lactose carrier has suggested that the functional unit of the carrier is a monomer,[78] and thus earlier complex models[79] based on a dimeric carrier lack a physical basis. However, in one major respect, the kinetics of the carrier fall outside the realm of simple rapid equilibrium binding models: in the case of lactose, an imposed $\Delta\bar{\mu}_H$ elicits an apparent increase in affinity of up to 200-fold. While the absence of such an affinity change in the case of GalSGal implies that affinity changes cannot be central to the gradient-coupling mechanism,[77] it remains to be seen whether models in which the rapid equilibrium assumption has been abandoned will yield a satisfactory explanation of these data, or whether additional postulates concerning carrier regulation will be needed.

E. Random-Binding Models

Analysis of nonequilibrium *random* binding models[80] (Figure 4C and D) has demonstrated that single or multiple michaelian kinetics can result with respect to $[S]_o$. Algebraic analysis of the model predicts that for a symporter, as $[H^+]_o$ is raised toward saturation, the multiple phase kinetics progressively reduce to a single michaelian function. The model was shown[80] to provide an excellent description of data[81] on H^+-sugar symport in *Chlorella* over a wide range of $[sugar]_o$. As anticipated by the model, transport is described as a single michaelian function, with low K_m, at low pH_o ($= 6.1$), and as a single function with high K_m at high pH (>8.4). In the intermediate pH range, two apparently discrete functions are generated.

In order to give a complete picture of the effects of $\Delta\bar{\mu}_H$ components on gradient-coupled transport, it will be necessary in the future to undertake detailed studies on the effects of ligands on the I-V curve of the transport system. A preliminary report of such data has been published[82] and a more complete analysis will appear soon.[83]

V. CONCLUDING REMARKS

In this short review, I have attempted to demonstrate the great utility of reaction kinetic models for transport systems. These models are highly flexible kinetically, and in a sense, constitute a class of minimal models for transport, which should be rigorously tested before other models are applied.

A high priority for research in the next few years should be the application of reaction kinetic models to H^+ transport systems in energy coupling membranes. This will necessitate a conceptual change in experimental design: instead of viewing the kinetics of H^+ translocation simply in terms of the overall driving force on the proton, it will be necessary to carefully consider the role of each component of the driving force as a kinetic determinant in its own right.

ACKNOWLEDGMENTS

I am grateful to Mick Hopgood for drawing the figures. Financial support was provided by the Agricultural and Food Research Council (U.K.).

REFERENCES

1. **Hilpert, W. and Dimroth, P.,** Conversion of the chemical energy of methyl-malonyl-CoA decarboxylation into an Na^+ gradient, *Nature (London)*, 296, 584, 1982.
2. **Schobert, B. and Lanyi, J. K.,** Halorhodopsin is a light-driven chloride pump, *J. Biol. Chem.*, 257, 10306, 1982.

3. **Mitchell, P.,** *Chemiosmotic Coupling in Oxidative and Photosynthetic Phosphorylation,* Glynn Research Ltd., Bodmin, England, 1966.

4. **Greville, G. D.,** A scrutiny of Mitchell's chemiosmotic hypothesis of respiratory chain and photosynthetic phosphorylation, *Curr. Top. Bioenerget.,* 3, 1, 1969.

5. **Harold, F. M.,** Membranes and energy transduction in bacteria, *Curr. Top. Bioenerget.,* 6, 83, 1977.

6. **Maloney, P. C.,** Energy coupling to ATP synthesis by the proton-translocating ATPase, *J. Membr. Biol.,* 67, 1, 1982.

7. **Goffeau, A. and Slayman, C. W.,** The proton-translocating ATPase of the fungal plasma membrane, *Biochim. Biophys. Acta,* 639, 197, 1981.

8. **Läuger, P.,** Thermodynamic and kinetic properties of electrogenic ion pumps, *Biochim. Biophys. Acta,* 779, 307, 1984.

9. **Läuger, P.,** Kinetic properties of ion carriers and channels, *J. Membr. Biol.,* 57, 163, 1980.

10. **Eisenbach, M. and Caplan, S. R.,** The light-driven proton pump of *Halobacterium halobium:* mechanism and function, *Curr. Top. Membr. Transport,* 12, 165, 1979.

11. **Crofts, A. R. and Wraight, C. A.,** The electrochemical domain of photosynthesis, *Biochim. Biophys. Acta,* 726, 149, 1983.

12. **Hansen, U.-P., Tittor, J., and Gradmann, D.,** Interpretation of current-voltage relationships for "active" ion transport systems. II. Nonsteady-state reaction kinetic analysis of class-I mechanisms with one slow time-constant, *J. Membr. Biol.,* 75, 141, 1983.

13. **Jagendorf, A. T. and Uribe, E.,** ATP formation caused by acid-base transition of spinach chloroplasts, *Proc. Natl. Acad. Sci. U.S.A.,* 55, 170, 1966.

14. **Reid, R. A., Moyle, J., and Mitchell, P.,** Synthesis of adenosine triphosphate by a protonmotive force in rat liver mitochondria, *Nature (London),* 212, 257, 1966.

15. **Maloney, P. C., Kashket, E. R., and Wilson, T. H.,** A protonmotive force drives ATP synthesis in bacteria, *Proc. Natl. Acad. Sci. U.S.A.,* 71, 3896, 1974.

16. **Westerhoff, H. V., Melandri, B. A., Venturoli, G., Azzone, G. F., and Kell, D. B.,** Mosaic protonic coupling hypothesis for free energy transduction, *FEBS Lett.,* 165, 1, 1984.

17. **Westerhoff, H. V., Melandri, B. A., Venturoli, G., Azzone, G. F., and Kell, D. B.,** A minimal hypothesis for membrane-linked free-energy transduction, *Biochim. Biophys. Acta,* 768, 257, 1984.

18. **Ferguson, S. J.,** Fully delocalized chemiosmotic or localized proton flow pathways in energy coupling? A scrutiny of the experimental evidence, *Biochim. Biophys. Acta,* 811, 47, 1985.

19. **Azzone, G. F., Petronilli, V., and Zoratti, M.,** "Cross-talk" between redox- and ATP-driven H^+ pumps, *Biochem. Soc. Trans.,* 12, 414, 1984.

20. **Sanders, D. and Hansen, U.-P.,** Mechanism of Cl^- transport at the plasma membrane of *Chara corallina.* II. Transinhibition and the determination of H^+/Cl^- binding order from a reaction kinetic model, *J. Membr. Biol.,* 58, 139, 1981.

21. **Grubmeyer, C., Cross, R. L., and Penefsky, H. S.,** Mechanism of ATP hydrolysis by beef heart mitochondrial ATPase. Rate constants for elementary steps in catalysis at a single site, *J. Biol. Chem.,* 257, 12092, 1982.

22. **Cross, R. L., Grubmeyer, C., and Penefsky, H. S.,** Mechanism of ATP hydrolysis by beef heart mitochondrial ATPase. Rate enhancements resulting from cooperative interactions between multiple catalytic sites, *J. Biol. Chem.,* 257, 12101, 1982.

23. **Hansen, U.-P., Gradmann, D., Sanders, D., and Slayman, C. L.,** Interpretation of current-voltage relationships for "active" ion transport systems. I. Steady-state reaction-kinetic analysis of Class-I mechanisms, *J. Membr. Biol.,* 63, 165, 1981.

24. **Maloney, P. C. and Schattschneider, S.,** Voltage sensitivity of the proton-translocating adenosine 5′-triphosphatase in *Streptococcus lactis, FEBS Lett.,* 110, 337, 1980.

25. **Cotton, N. P. J., Clark, A. J., and Jackson, J. B.,** Changes in membrane ionic conductance, but not changes in slip, can account for the non-linear dependence of the electrochemical proton gradient upon the electron-transport rate in chromatophores, *Eur. J. Biochem.,* 142, 193, 1984.

26. **O'Shea, P. S., Petrone, G., Casey, R. P., and Azzi, A.,** The current-voltage relationships of liposomes and mitochondria, *Biochem. J.,* 219, 719, 1984.

27. **Scholes, P. and Mitchell, P.,** Acid-base titration across the plasma membrane of *Micrococcus denitrificans:* factors affecting the effective proton conductance and respiratory rate, *J. Bioenerg.,* 1, 61, 1970.

28. **Maloney, P. C.,** Membrane H^+ conductance of *Streptococcus lactis, J. Bacteriol.,* 140, 197, 1979.

29. **Gradmann, D., Hansen, U.-P., and Slayman, C. L.,** Reaction-kinetic analysis of current-voltage relationships for electrogenic pumps in *Neurospora* and *Acetabularia, Curr. Top. Membr. Trans.,* 16, 257, 1982.

30. **Gadsby, D. C., Kimura, J., and Noma, A.,** Voltage dependence of Na/K pump current in isolated heart cells, *Nature (London),* 315, 63, 1985.

31. **Spanswick, R. M.,** Evidence for an electrogenic ion pump in *Nitella translucens.* I. The effects of pH, K$^+$, Na$^+$, light and temperature on the membrane potential and resistance, *Biochim. Biophys. Acta,* 288, 73, 1972.

32. **Walker, N. A. and Smith, F. A.,** The H$^+$ ATPase of the *Chara* cell membrane: its role in determining membrane p.d. and cytoplasmic pH, in *Transmembrane Ionic Exchanges in Plants,* Thellier, M., Monnier, A., Demarty, M., and Dainty, J., Eds., Centre de Neurochimie du Recherche, Paris, 1977, 255.

33. **Spanswick, R. M.,** Biophysical control of electrogenicity in the characeae, in *Plant Membrane Transport: Current Conceptual Issues,* Spanswick, R. M., Lucas, W. J., and Dainty, J., Eds., Elsevier, Amsterdam, 1980, 305.

34. **Beilby, M. J.,** Current-voltage characteristics of the proton pump at *Chara* plasmalemma. I. pH dependence, *J. Membr. Biol.,* 81, 113, 1984.

35. **Lucas, W. J.,** Mechanism of acquisition of exogenous bicarbonate by internodal cells of *Chara corallina, Planta,* 156, 181, 1982.

36. **Woelders, H., van der Zande, W. J., Colen, A.-M. A. F., Wanders, R. J. A., and van Dam, K.,** The phosphate potential maintained by mitochondria in State 4 is proportional to the proton-motive force, *FEBS Lett.,* 179, 278, 1985.

37. **Maloney, P. C. and Hansen, F. C.,** Stoichiometry of proton movements coupled to ATP synthesis driven by a pH gradient in *Streptococcus lactis, J. Membr. Biol.,* 66, 63, 1982.

38. **Sanders, D., Hansen, U.-P., Gradmann, D., and Slayman, C. L.,** Generalized kinetic analysis of ion-driven cotransport systems: a unified interpretation of selective ionic effects on Michaelis parameters, *J. Membr. Biol.,* 77, 123, 1984.

39. **Schindler, H. and Nelson, N.,** Proteolipid of adenosinetriphosphatase from yeast mitochondria form proton-selective channels in planar lipid bilayers, *Biochemistry,* 21, 5787, 1982.

40. **Patlak, C. S.,** Contributions to the theory of active transport. II. The gate type non-carrier mechanism and generalizations concerning tracer flow, efficiency and measurement of energy expenditure, *Bull. Math. Biophys.,* 19, 209, 1957.

41. **Chapman, J. B., Johnson, E. A., and Kootsey, J. M.,** Electrical and biochemical properties of an enzyme model of the sodium pump, *J. Membr. Biol.,* 74, 139, 1983.

42. **Slayman, C. L.,** Plasma membrane proton pumps in plants and fungi, *Bio-Science,* 35, 34, 1985.

43. **Ahlers, J.,** Temperature effects on kinetic properties of plasma membrane ATPase from the yeast *Saccharomyces cerevisiae, Biochim. Biophys. Acta,* 649, 550, 1981.

44. **Brooker, R. J. and Slayman, C. W.,** Effects of Mg^{2+} ions on the plasma membrane [H$^+$]-ATPase of *Neurospora crassa.* II. Kinetic studies, *J. Biol. Chem.,* 258, 8833, 1983.

45. **Gradmann, D., Hansen, U.-P., Long, W. S., Slayman, C. L., and Warncke, J.,** Current-voltage relationships for the plasma membrane and its principal electrogenic pump in *Neurospora crassa.* I. Steady-state conditions, *J. Membr. Biol.,* 39, 333, 1978.

46. **Kuroda, H., Warncke, J., Sanders, D., Hansen, U.-P., Allen, K. E., and Bowman, B. J.,** Effects of vanadate on the electrogenic proton pump in *Neurospora,* in *Plant Membrane Transport: Current Conceptual Issues,* Spanswick, R. M., Lucas, W. J., and Dainty, J., Eds., Elsevier, Amsterdam, 1980, 507.

47. **Slayman, C. L. and Slayman, C. W.,** The electrogenic proton pump in *Neurospora crassa,* in *Chemiosmotic Proton Circuits in Biological Membranes,* Skulachev, V. P. and Hinckle, P. C., Eds., Addison-Wesley, Reading, Mass., 1981, 337.

48. **Sanders, D., Hansen, U.-P., and Slayman, C. L.,** Role of the plasma membrane proton pump in pH regulation in non-animal cells, *Proc. Natl. Acad. Sci. U.S.A.,* 78, 5903, 1981.

49. **Slayman, C. L. and Sanders, D.,** Electrical kinetics of proton pumping in *Neurospora,* in *Electrogenic Transport: Fundamental Principles and Physiological Implications,* Blaustein, M. P. and Lieberman, M., Eds., Raven Press, New York, 1984, 307.

50. **Takeuchi, Y., Kishimoto, U., Ohkawa, T., and Kami-ike, N.,** A kinetic analysis of the electrogenic pump of *Chara corallina.* II. Dependence of pump activity on external pH, *J. Membr. Biol.,* 86, 17, 1985.

51. **Slayman, C. L. and Sanders, D.,** pH dependence of proton pumping in *Neurospora,* in *Hydrogen Ion Transport in Epithelia,* Forte, J. G., Warnock, D. G., and Rector, F. C., Eds., Wiley Interscience, New York, 1984, 47.

52. **Slayman, C. L. and Sanders, D.,** Steady-state kinetic analysis of an electroenzyme, in *The Molecular Basis of Movement Through Membranes,* Quinn, P. J. and Pasternak, C. A., Eds., Biochemical Society, London, 1985, 11.

53. **Bowman, B. J., Blasco, F., and Slayman, C. W.,** Purification and characterization of the plasma membrane ATPase of *Neurospora crassa, J. Biol. Chem.,* 256, 12343, 1981.

54. **Peters, P. H. J. and Borst-Pauwels, G. W. F. H.,** Properties of the plasma membrane ATPase and mitochondrial ATPase of *Saccharomyces cerevisiae, Physiol. Plant.,* 46, 330, 1979.

55. **Crane, R. K., Miller, D., and Bihler, I.,** The restrictions on possible mechanisms of intestinal active transport of sugars, in *Membrane Transport and Metabolism,* Kleinzeller, A. and Kotyk, A., Eds., Academic Press, New York, 1961, 439.

56. **Schultz, S. G. and Curran, P. F.,** Coupled transport of sodium and organic solutes, *Physiol. Rev.,* 50, 637, 1970.

57. **Heinz, E., Geck, P., and Wilbrandt, W.,** Coupling in secondary active transport. Activation of transport by cotransport and/or countertransport with the fluxes of other solutes, *Biochim. Biophys. Acta,* 255, 442, 1972.

58. **Geck, P. and Heinz, E.,** Coupling in secondary transport. Effect of electrical potentials on the kinetics of ion-linked cotransport, *Biochim. Biophys. Acta,* 443, 49, 1976.

59. **West, I. C.,** Energy coupling in secondary active transport, *Biochim. Biophys. Acta,* 604, 91, 1980.

60. **Hopfer, U. and Groseclose, R.,** The mechanism of Na^+-dependent D-glucose transport, *J. Biol. Chem.,* 255, 4453, 1980.

61. **Harrison, D. A., Rowe, G. W., Lumsden, C. J., and Silverman, M.,** Computational analysis of models for cotransport, *Biochim. Biophys. Acta,* 774, 1, 1984.

62. **Cuppoletti, J. and Segel, I. H.,** Kinetic analysis of active membrane transport systems: equations for net velocity and isotope exchange, *J. Theor. Biol.,* 53, 125, 1975.

63. **Cuppoletti, J. and Segel, I. H.,** Kinetics of sulfate transport by *Penicillium notatum.* Interactions of protons, sulfate and calcium, *Biochemistry,* 14, 4712, 1975.

64. **Stock, J. and Roseman, S.,** A sodium-dependent sugar co-transport system in bacteria, *Biochem. Biophys. Res. Commun.,* 44, 132, 1971.

65. **Niiya, S., Moriyama, Y., Futai, M., and Tsuchiya, T.,** Cation coupling to melibiose transport in *Salmonella typhimurium, J. Bacteriol.,* 144, 192, 1980.

66. **Page, M. G. P. and West, I. C.,** The kinetics of the β-galactoside-proton symport of *Escherichia coli, Biochem. J.,* 196, 721, 1981.

67. **Komor, E. and Tanner, W.,** The hexose-proton cotransport system of *Chlorella.* pH-dependent change in K_m values and translocation constants of the uptake system, *J. Gen. Physiol.,* 64, 568, 1974.

68. **Schwab, W. G. W. and Komor, E.,** A possible mechanistic role of the membrane potential in proton-sugar cotransport of *Chlorella, FEBS Lett.,* 87, 157, 1978.

69. **Le Blanc, G., Rimon, G., and Kaback, H. R.,** Glucose 6-phosphate transport in membrane vesicles isolated from *Escherichia coli:* effect of imposed electrical potential and pH gradient, *Biochemistry,* 19, 2522, 1980.

70. **ten Brink, B., Otto, R., Hansen, U.-P., and Konings, W. N.,** Energy recycling by lactate efflux in growing and nongrowing cells of *Streptococcus cremoris, J. Bacteriol.,* 162, 383, 1985.

71. **Hansen, U.-P. and Slayman, C. L.,** Current-voltage relationships for a clearly electrogenic cotransport system, in *Membrane Transport Processes,* Vol. 1, Hoffman, J. F., Ed., Raven Press, New York, 1978, 141.

72. **Sanders, D., Slayman, C. L., and Pall, M. L.,** Stoichiometry of H^+/amino acid cotransport in *Neurospora crassa* revealed by current-voltage analysis, *Biochim. Biophys. Acta,* 735, 67, 1983.

73. **Slayman, C. L., Hansen, U.-P., Gradman, D., and Sanders, D.,** Kinetic modelling of electrophoretic cotransport, in *Membrane Permeability: Experiments and Models,* Bretag, A. H., Ed., Techsearch, Adelaide, Australia, 1983, 3.

74. **Eddy, A. A.,** Slip and leak models of gradient-coupled transport, *Trans. Biochem. Soc. London,* 8, 271, 1980.

75. **Ahmed, S. and Booth, I. R.,** The effects of partial and selective reduction in the components of the proton-motive force on lactose uptake in *Escherichia coli, Biochem. J.,* 200, 563, 1981.

76. **Overath, P. and Wright, J. K.,** Lactose permease: a carrier on the move, *Trends Biochem. Sci.,* 8, 404, 1983.

77. **Wright, J. K., Dornmair, K., Mitaku, S., Moroy, T., Neuhaus, J. M., Seckler, R., Vogel, H., Weigel, U., Jähning, F., and Overath, P.,** Lactose:H^+ carrier of *Escherichia coli:* kinetic mechanism, purification, and structure, *Ann. N.Y. Acad. Sci.,* 456, 326, 1985.

78. **Wright, J. K., Weigel, U., Lustig, A., Bocklage, H., Mieschendahl, M., Müller-Hill, B., and Overath, P.,** Does the lactose carrier of *Escherichia coli* function as a monomer?, *FEBS Lett.,* 162, 11, 1983.

79. **Lombardi, F. J.,** Lactose-$H^+(OH^-)$ transport system of *Escherichia coli.* Multistate gated pore model based on half sites stoichiometry for high-affinity substrate binding in a symmetrical dimer, *Biochim. Biophys. Acta,* 649, 661, 1981.

80. **Sanders, D.,** Generalized kinetic analysis of ion-driven cotransport systems. II. Random ligand binding as a simple explanation for non-michaelian kinetics, *J. Membr. Biol.,* 90, 67, 1986.

81. **Komor, E. and Tanner, W.,** Simulation of a high- and low-affinity sugar-uptake system in *Chlorella* by a pH-dependent change in the K_m of the uptake system, *Planta,* 123, 195, 1975.

82. **Blatt, M. R. and Slayman, C. L.,** "Active" K^+ transport in *Neurospora:* cotransport with H^+, *Plant Physiol.,* 75(Suppl.), 183, 1984.

83. **Blatt, M. R. and Slayman, C. L.,** in preparation.

Chapter 4

TEMPERATURE EFFECTS ON BACTERIAL GROWTH RATES

T. A. McMeekin, June Olley, and D. A. Ratkowsky

TABLE OF CONTENTS

I. INTRODUCTION

Bacterial growth is defined as the coordinated summation of a complex array of processes including chemical synthesis, assembly, polymerization, biosynthesis, fueling and transport, the consequence of which is the production of new cells.[1]

The rate at which growth occurs is affected by many factors, including temperature. Temperature relations provide an excellent example of bacterial diversity and it has been suggested that bacterial growth is possible only if liquid water is present.[2] Thus, at the lower end of the scale microbial growth is usually limited by freezing of the growth medium, an effect often attributed to lowered water activity[3] rather than temperature per se. The upper limit for growth is currently the subject of much debate. It is well established that growth can occur at temperatures in excess of 100°C,[4] while recent observations[5,6] indicate growth of bacteria at temperatures of more than 300°C in superheated waters kept liquid by pressures in excess of 260 atm. These observations have been questioned.[7,8]

While different bacteria grow over a wide range of temperatures, growth of individual organisms is usually restricted to a range of approximately 40°C and arbitrary temperature categories have been suggested based on the minimum, optimum, and maximum temperatures for growth. Definitions of these categories vary from author to author. In the following discussion we define psychrophiles, psychrotrophs, mesophiles, and thermophiles in terms of their cardinal temperatures and a representative of each category is presented in Table 1.

For all bacteria there are minimum and maximum temperatures beyond which growth does not occur. The temperature at which the fastest rate of growth is observed is termed the optimum growth temperature (T_{opt}).

The purpose of this chapter is to examine mathematical models describing the effect of temperature on bacterial growth and to examine how this information may be used to predict the extent of microbial growth on various substrates.

II. EFFECT OF TEMPERATURE ON BACTERIAL GROWTH

If a bacterial culture is grown in a liquid medium in which nutrients are nonlimiting, growth is said to be unrestricted.[1] In this mode the rate of increase of bacterial cells in a culture is proportional only to the number of cells present at a particular time. Mathematically,

$$dN/dt = kN \tag{1}$$

where N is the concentration of bacterial cells (number per unit volume), t is the time, and the constant of proportionality k is called the specific growth rate constant and is an index of growth rate for a particular organism. Alternatively, growth can be described as following an exponential law.

In practice, for a given bacterial culture, the same values of k tend to be obtained whether cell concentration, bacterial mass, protein, DNA, or any other "extensive property of the growing system" is measured.[9]

The specific growth rate constant (k) is a constant only for a single temperature. It has a maximum value at the optimum temperature for growth and is zero at temperatures in excess of the maximum temperature or less than the minimum temperature for growth. Typically, growth rate declines rapidly at temperatures greater than optimum.

The effect of temperature on the rate of microbial growth has traditionally been described by a modification of the Arrhenius equation in which microbiologists have simply substituted

Table 1
CARDINAL TEMPERATURES (KELVIN) FOR
FOUR SELECTED ORGANISMS

Class	Organism	T_{min}	T_{opt}	T_{max}
Psychrophile	Antarctic strain 755	251	288	293
Psychrotroph	*Pseudomonas* sp.	263	300	310
Mesophile	*Serratia marcescens*	278	308	314
Thermophile	*Bacillus stearothermophilus*	308	337	349

the growth rate constant of bacterial growth for the rate constant of a simple chemical reaction and employed the expression:

$$d\log k/dT = E/RT^2 \qquad (2)$$

where T is the absolute temperature (degrees Kelvin), R is the universal gas constant (8.314 J/mol/deg), and E is an empirically determined quantity called the activation energy. Under the assumption that the activation energy is not a function of temperature, Equation 2 integrates to yield:

$$k = A \exp(-E/RT) \qquad (3)$$

where A is the "collision" or "frequency" factor. Equation 3 is generally known as the Arrhenius law and this expression has had some notable success in describing the temperature dependence of many chemical reactions. When Equation 3 is valid, a plot of log k against 1/T gives a straight line. Nevertheless, even with simple reactions, the assumed constancy of E with temperature is only an approximation, and studies involving temperatures ranging over more than approximately 30°C reveal that activation energies tend to decrease with increased temperature, rather than remaining constant.

Bacterial growth involves the interaction of a highly complex series of reactions comprising both catabolic and anabolic processes and it is therefore not surprising that Arrhenius plots of bacterial specific growth rate deviate markedly from linearity. Some workers[1] invoke a linear response in the mid-range of temperature where normal chemical kinetics seem to apply, but at higher and lower temperatures the specific growth rate is less than the value predicted by extrapolation of the Arrhenius equation. More complex relationships involving "broken" Arrhenius plots have also been suggested.[10,11] The authors[12] interpret the response to temperature to be a continuously downward sloping curve throughout the entire suboptimum temperature range. Figure 1, which represents five bacteria and a mold, was redrawn from Johnson et al.[13] and from this Ratkowsky et al.[12] concluded "that the data do not even remotely approximate a straight line relationship at any portion of the range".

III. THE SQUARE ROOT MODEL

A. Development of the Model

Although the Arrhenius equation does not accurately describe the effect of temperature on the rate of bacterial growth, no alternative models were suggested until the work of Ratkowsky et al.[12,14] The models described in these papers had their genesis in attempts to predict the spoilage of fish at various temperatures.[15,16]

Ohta and Hirahara[17] noted that a plot of the square root of the rate of breakdown of nucleotides in carp muscle vs. temperature was nearly linear. Because it is often claimed that rigor mortis precedes microbiological spoilage, this observation prompted an examination

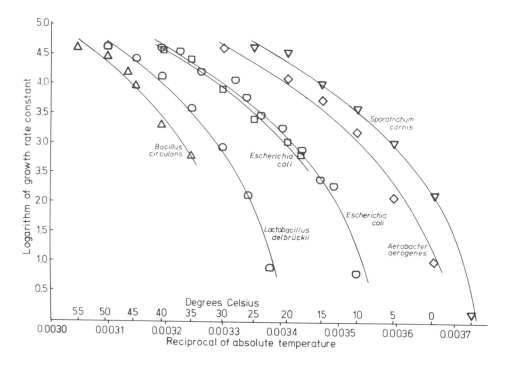

FIGURE 1.   Arrhenius plot of six sets of data redrawn from Johnson et al.[13] The solid curves correspond to the equation $\sqrt{k} = b(T - T_{min})$. (From Ratkowsky, D. A., Olley, J., McMeekin, T. A., and Ball, A., *J. Bacteriol.*, 149, 1, 1982. With permission.)

of the effects of temperature on bacterial growth. Ratkowsky et al.[12] demonstrated that the square root relationship could be used to model the temperature dependence of the specific growth rate constant at temperatures between the minimum and optimum temperatures for growth.

The relationship between temperature and specific growth rate can be written as follows:

$$\sqrt{k} = b(T - T_{min}) \tag{4}$$

where k is the specific growth rate described in Section II, b is the slope of the regression line, T is the temperature (degrees Kelvin), and T_{min} is the temperature where the regression line cuts the temperature axis. In Ratkowsky et al.,[12] T_{min} was denoted as T_o.

Equation 4 was applied to more than 50 sets of data by Ratkowsky et al.[12] and it also accurately described the minimum to optimum temperature range for an additional 30 organisms studied by Ratkowsky et al.[14] Data sets in the latter paper include square root plots based on the original data of Mohr and Krawiec[11] and Reichardt and Morita.[10] Thus, the square root model also accurately describes the effect of temperature on the specific growth rate of organisms which appeared to have two distinct suboptimal temperature characteristics based on Arrhenius plots.

The data sets presented by Ratkowsky et al.[12,14] include representatives of the categories psychrophile, psychrotroph, mesophile, and thermophile. It is also of interest to note that Equation 4 describes the effect of temperature on growth rate of "black smoker" bacteria[6] at 150, 200, and 250°C and gives a T_{min} value of 110°C.

Naturally, Equation 4 cannot be applied to temperatures close to or greater than the optimum temperature for growth. As is obvious from Arrhenius plots, the specific growth rate declines very markedly at superoptimal temperatures where considerable heat denatur-

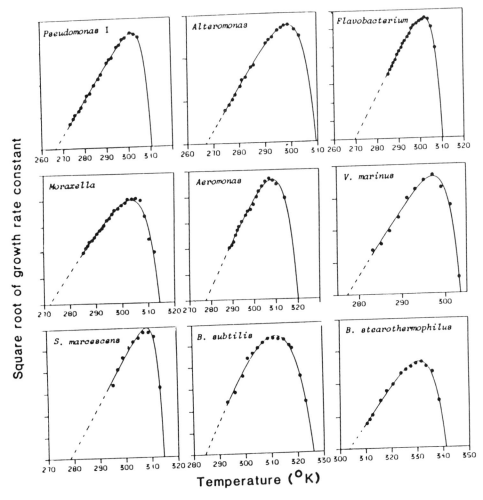

FIGURE 2. Data and fitted lines for Equation 5 for nine bacterial cultures. (From Ratkowsky, D. A., et al., *J. Bacteriol.*, 154, 1222, 1983. With permission.)

ation of the cell protein occurs, and death may result with little more than a 5°C rise in temperature.

To describe the temperature dependence of growth rate in this region Ratkowsky et al.[14] extended Equation 4 to cover the entire temperature range of growth. The empirical nonlinear regression model is

$$\sqrt{k} = b(T - T_{min}) \{1 - \exp[c(T - T_{max})]\} \tag{5}$$

where b, c, T_{min}, and T_{max} are the four parameters to be estimated. In Equation 5, T_{min} together with T_{max} represent the points at which the regression line intersects the temperature axis. Ratkowsky et al.[14] applied Equation 5 to 30 data sets, including 16 from new isolates, 12 from Mohr and Krawiec[11] and 2 from Reichardt and Morita.[10] Typical examples are shown in Figure 2.

When T is much lower than T_{max}, the contribution of the term in braces is negligible and Equation 5 reduces to Equation 4, the original square root model. As T increases to approach T_{max}, the term becomes increasingly more important until eventually it dominates and the growth rate falls as T exceeds T_{opt}, reaching zero when $T = T_{max}$.

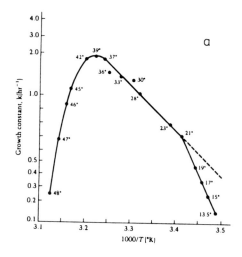

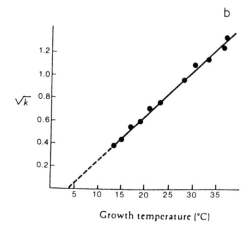

FIGURE 3.   (a) Arrhenius plot of growth rate constant of *Escherichia coli*. (b) Plot of data from Figure 3a according to Equation 4. Numbers against data points are temperatures in °C. (From Ingraham, J. L., Maaløe, O., and Neidhardt, F. C., *Growth of the Bacterial Cell*, Sinauer Associates, Sunderland, Mass., 1983. With permission.)

B. Cardinal Temperatures

Use of the square root model (Equation 5) allows estimation of the parameters T_{min} and T_{max} and of the optimum growth temperature for the organism. Estimates of the cardinal temperatures so obtained for four bacteria are given in Table 1. The selected organisms are representative of the categories psychrophile, psychrotroph, mesophile, and thermophile; a more complete listing of cardinal temperature is presented by Ratkowsky et al.[12,14]

T_{min} and T_{max} occur at the points where the square root plot intersects the temperature axis, i.e., at both points the growth rate is zero. As it is difficult to obtain accurate data at very low growth rates, it must be emphasized that T_{min} and T_{max} are notional temperatures. A further reason why T_{min} is a notional temperature is that for psychrotrophs and psychrophiles, T_{min} is below the freezing point of the substrate. This would alter the water activity and inhibit growth. T_{min} values could be realized for mesophiles and thermophiles.

Ingraham et al.[1] indicated that the T_{min} value of *Escherichia coli* (3.5°C) does not accurately predict the minimum temperature for growth (8°C). This suggests that the square root relationship does not hold in the extreme low range. Inspection of their data (Figure 3) indicates almost perfect linearity between 13.5 and 35°C. Shaw et al.[18] found no growth for *E. coli* at temperatures less than 7.8°C. In this laboratory we have studied the effect of temperature on the growth rate of a strain of *E. coli* between 2 and 20°C. The minimum observed temperature for growth was also 7.8°C; 12% turbidity was reached in 16 days. No growth occurred at 7.2°C after 30 days incubation. Similarly, cells maintained at 5°C for 5 months failed to grow. Therefore, although the actual minimum temperature for growth may occur several degrees above the extrapolated T_{min} value, our data indicated that the square root relationship remains valid to the point at which growth ceases. Despite the fact that T_{min} values may be considered conceptual temperatures there are good reasons for their determination (see Section V).

Estimates of T_{min} vary from 248K for a psychrophile to 296K for a thermophile.[12] Psychrotrophs have T_{min} values of 260 to 270K with a gradation to the mesophilic category (270 to 280K). Our observations[12,19] suggest a distinct break in T_{min} values between psychrophiles and psychrotrophs, but this is probably due to insufficient strains having been studied. Although aberrant values were computed[14] for *Vibrio psychroerythrus* and *V. marinus* which had psychrophilic T_{opt} and T_{max} values, we suggested that T_{min} values may provide a a useful guide to categorize bacteria in relation to temperature. Preliminary studies have begun in our laboratories to isolate bacteria from Antarctic saline lakes. This may yield salt tolerant, psychrophilic, or psychrotrophic bacteria useful for studies of growth rates at subzero temperatures but unaffected by lowered water activity.[19] T_{min} values are also necessary to convert data to relative rate curves for incorporation into temperature function integrators. This will be discussed in Section V.

C. Effect of the Lag Phase on the Square Root Model

In the work described by Ratkowsky et al.,[12,14] the experimental conditions involved inoculation of an exponential phase culture into a fresh medium of the same composition. This procedure eliminated the presence of a lag phase and allowed unrestricted growth at a rate characteristic of each temperature. Chandler and McMeekin[20] examined the hypothesis that the effect of temperature on the lag phase of growth might also be described by the square root model. This arose from practical considerations in monitoring psychrotrophic bacterial spoilage of pasteurized milk where a lag phase might be expected to occur. A pseudomonad isolated from milk was grown through to the stationary phase of growth for 96 hr at 25°C. Measurement of the effect of temperature on length of lag phase showed that it was described by a square root relationship and a T_{min} value similar to that obtained with a culture grown exponentially (267.3 and 267.9K, respectively).

Smith[21] has also determined the effect of temperature on the lag and generation times of *E. coli*. Analysis of the data indicates that both parameters obey the square root model and yield similar T_{min} values (274.9 and 275.8K).

IV. RECONCILING THE ARRHENIUS LAW AND THE SQUARE ROOT MODEL

We now turn our attention to the suboptimal temperature region of growth to see if there can be any reconciliation between the "square root" model, given by Equation 4 and the Arrhenius law. We have already remarked that the activation energy E, even for chemical reactions, is usually not constant, but instead tends to decrease with increasing temperature. Thus, the nonapplicability of the Arrhenius law to the modeling of the temperature dependence of microbiological growth can be viewed as resulting from the value of E changing with temperature. We will now present some mathematics which show how E must change with temperature to give rise to a square root model.

Table 2
ACTIVATION ENERGY (E) EXPRESSED AS A
FUNCTION OF T − T_{min} FOR FOUR SELECTED
BACTERIA

$E = 2RT^2/(T - T_{min})$, kJ/mol

T − T_{min}	Psychrophile $T_{min} = 251$	Psychrotroph $T_{min} = 263$	Mesophile $T_{min} = 278$	Thermophile $T_{min} = 308$
5	218	239	266	326
10	113	124	138	168
15	78.4	85.8	95.0	115
20	61.1	66.5	73.6	89.5
25	50.7	55.2	61.1	73.6
30	43.8	47.7	52.7	63.2

Taking the logarithm of both sides of Equation 4 and differentiating with respect to temperature yields:

$$dlog\ k/dT = 2/(T - T_{min}) \qquad (6)$$

However, the differential form of the Arrhenius Law (Equation 2) gives this derivative as:

$$dlog\ k/dT = E/RT^2 \qquad (7)$$

Hence, it follows that the activation energy must be related to temperature by the expression:

$$E = 2RT^2/(T - T_{min}) \qquad (8)$$

Table 2 gives some typical values of E for each of four bacteria, the same ones as in Table 1, representing psychrophiles, psychrotrophs, mesophiles, and thermophiles. Temperature differences T − T_{min} range from 5 to 30°C in Table 2. For a given organism, the rate of change of E is greater for low values of T − T_{min} than for higher values. A high activation energy means a high energy barrier to reaction so the rate of bacterial growth will be low for organisms that are only 10°C or so above their T_{min} values. For thermophiles, the barrier is greater than for psychrophiles and these organisms will not show appreciable growth until they are at temperatures of approximately 15 to 20°C above their T_{min} values. Table 2 is calculated up to a T − T_{min} of 30°C. It should be borne in mind that some organisms have a T_{opt} − T_{min} of less than 30°C.

It can be seen how the concept has arisen of an approximately constant activation energy in the growth temperature range of a microorganism characterized as "normal" or "linear" by Ingraham et al.[1] A 5° temperature difference 20 to 25°C above T_{min} only causes a 17% change in activation energy, while a 5° temperature change 5 to 10°C above T_{min} causes a 100% change in activation energy. Marr et al.[22] and Shaw[23] discuss shifting a bacterium or yeast into or out of the temperature range, which they have called one of "constant" activation energy. The sharp changes in activation energy of *Escherichia coli* at 20°C (Figure 3a) have been attributed to the sudden loss of ability to produce β-galactosidase below this temperature. The good fit to a square root relationship (Figure 3b) suggests that this evidence was circumstantial and that there was no break in the linear square root relationship with temperature for *E. coli*, or, for that matter, any other microorganism. Further clues may be sought among other proteins which are hyperinduced or totally repressed with shifts in temperature. The content in *E. coli* of some of these proteins is linearly related to temperature and they are called "thermometer proteins".[24]

V. TEMPERATURE FUNCTION INTEGRATION

A. Application of the Square Root Model

The square root relationship (Equation 4) has been found to apply to the spoilage of foods as well as to the growth of individual bacteria and molds. For example, Pooni and Mead[25] tested several mathematical models on 28 sets of spoilage data from 14 published studies on poultry meat at temperatures between -2 and $25°C$ and found that Equation 4 was the most appropriate up to $15°C$.

Spoilage of foods such as meat and poultry is the result of bacterial activity on the food substrates. Psychrotrophs are capable of growing at refrigeration temperatures, thereby causing spoilage of chilled stored foods, although their temperature optima may be $25°C$ or greater (see Table 1). At storage temperatures close to $0°C$, the dominant spoilage organisms are the pseudomonads, having T_{min} values of the order of 263 to 266K. A study of 70 sets of spoilage data[15,16] indicated that most spoilage was due to psychrotrophic organisms. If one uses Equation 4 with a T_{min} value of approximately 263K, corresponding to the pseudomonad in Table 1, one can compare rates at any temperature T and $0°C$.

From Equation 4

$$(k/k_o)^{1/2} = (T - 263)/(273 - 263) \qquad (9)$$

where k_o is the specific growth rate at $0°C$. Using temperatures in degrees centigrade instead of degrees Kelvin Equation 9 can be rewritten as

$$(k/k_o)^{1/2} = 1 + 0.1t \qquad (10)$$

where t is the temperature in degrees centigrade, Figure 4 shows this equation in graphical form.

The relative rate curve thus obtained has been used to program electronic devices known as temperature function integrators similar to that described by Owen and Nesbitt.[26] In these devices the circuitry receives impulses from a thermistor and displays the integrated information as an equivalent number of days at an arbitrary reference temperature (usually $0°C$). Hence, the integrator is effectively a device which mimics psychrotrophic growth and displays a figure related to the amount of growth expected to occur during a particular time/temperature regime.

Integrators programmed with different relative rate curves based on mesophilic T_{min} values may also be used to monitor deterioration of products due to mesophilic bacteria, e.g, growth of *E. coli* in meat offals[21,27] and production of histamine in fish.[28]

In the psychrotrophic mode (i.e., based on a 263K curve) several workers have noted a deviation from the predicted curve at elevated temperatures.[20,25,29] They attribute these deviations to an increased contribution from mesophilic bacteria (with higher T_{min} values) as the temperature rises. From a practical viewpoint, this may not be a serious deviation as prolonged storage at such temperatures would lead to rapid deterioration and rejection of the product. However, transient or short excursions into the upper temperature range may have a significant effect on the overall rate of spoilage even if the product is returned to chill storage. If this is the case, the electronics of the integrator could be arranged to incorporate two relative rate curves, each operating over specific temperature ranges. In pasteurized, homogenized milk, pseudomonads dominate at storage temperatures up to $15°C$ with mesophilic, Gram positive and negative bacteria dominant at higher temperatures. Overall, bacterial growth is best simulated using a 264K curve up to $15°C$ and a 270K above $15°C$.[30]

Until recently, few attempts were made to predict the shelf life of aerobically stored foods

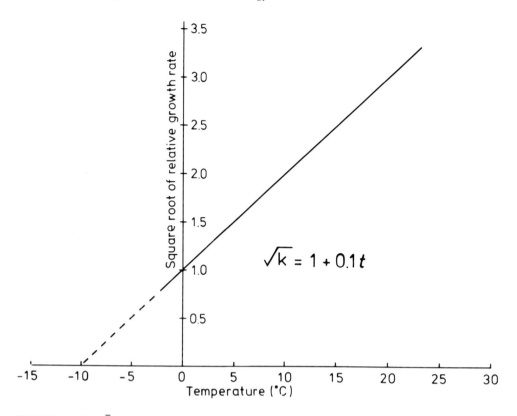

FIGURE 4.   The $\sqrt{k}$ relationship with temperature plotted as a rate relative to that at 0°C with a T_{min} of 263.16K (-10°C).

on the basis of temperature history, because bacterial counting was tedious and time consuming, allowing food to deteriorate before the completion of the experiment. Integrators display the amount of predicted shelf life already used up. Rapid automated conductance assays[31] of bacterial growth will enable the electronic and bacterial data to be compared more easily. The circuitry of the integrator can then be suitably modified, if necessary, to reflect more accurately the spoilage process.

B. Temperature Cycling

Temperature function integration provides a summation of time/temperature history expressed as equivalent time at a specified reference temperature. It assumes that the amount of spoilage resulting from the time a product spends at each temperature can be pooled to provide a composite result.

It has been shown in Section IV that changes in temperature out of the so-called "normal range" cause a delay in a bacterium or yeast stabilizing to its normal growth rate for the new temperature. However, these transient periods are usually of the order of two generation times.[22,23]

Figure 5 shows an example of temperature cycling on a natural product. The taste panel's average cooked flavor score of cod was the same whether the fish had been stored at 1°C, 15°C, or cycled between these temperatures for periods equivalent to equal numbers of days at 0°C.[32] A foodstuff warming up or cooling down gradually would probably be described by temperature function integration based on Equation 4. However, in a situation where extreme temperature fluctuations occurred it is conceivable that the temperature change might be too great for the bacterial enzymes to adapt immediately to the new temperature.

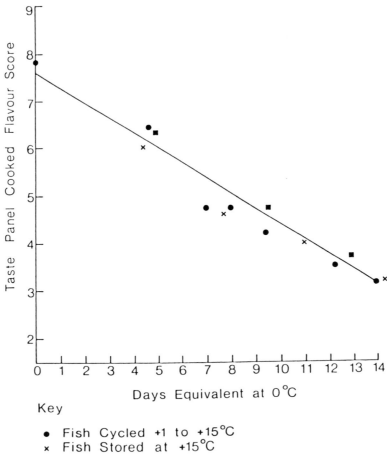

Key

● Fish Cycled +1 to +15°C
× Fish Stored at +15°C
■ Fish Stored at +1°C

FIGURE 5. Taste panel flavor scores of cod vs. equivalent days on ice. Fish cycled +1°C to +15°C, fish stored at +15°C, and fish stored at +1°C. (Unpublished data from the work of Spencer and Baines 1964, reappraised by R. M. Storey and D. Owen.)

Gibbs et al.[33] report initial studies with pure cultures in which the temperature was cycled between 4 and 15°C. In these experiments differences were found between the responses of the organism and that of the temperature function integrator. We presume that some of these organisms had T_o values different from the 263K value on which the integrator circuitry is based. However, the mean response, corresponding to mixed culture spoilage, agreed more closely with that of the integrator. Therefore, studies with the mixed microflora in foods are being actively pursued.

VI. EFFECT OF OTHER CONSTRAINTS ON BACTERIAL GROWTH

Equations 4 and 5 were developed from data in which temperature was the sole factor restricting the rate of growth but they are also valid when other constraints, such as lowered water activity or modified atmospheres, are placed on growth.

Doe[34] found that an extension of Equation 5, incorporating terms for water activity (a_w), fitted the data of Ayerst[35] on mold germination and growth at water activities ranging from 0.75 to 0.90 (Figure 6) and Olley and Doe[36] confirmed the validity of the square root relationship for growth of *Clostridium botulinum* at reduced water activities using the data of Ohye and Christian.[37]

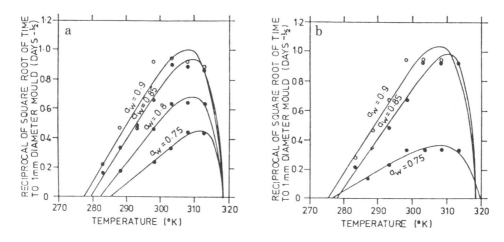

FIGURE 6.   (a) *Aspergillus chevalieri* var. *intermedius*. (b) *A. amstelodami*. Equation 5 applied to the data of Ayerst[35] for the growth of two fungi, at different water activities.

The combined effect of temperature and water activity on growth rate can be shown as a three-dimensional figure (Figure 7). Outside the bounds of the figure the temperature/water activity combinations prevent growth of the organism. Currently, we are examining the effect of these two factors on the growth rate of halotolerant and halophilic bacteria isolated from salted fish with the object of developing a model to specify safe drying and storage conditions.

Equation 4 also has been used to describe the effect of temperature on the deterioration of modified atmosphere and vacuum packed fish.[31]

Thomas[38] has pointed out that the data of Gill and Penney[39] on penetration of bacteria into meat appear to obey Equation 4, suggesting that bacterial proteolytic enzyme production at the higher temperatures may break down the connective tissue barrier to penetration of bacteria. Gill and Penney[39] found that the motility of the proteolytic strains was irrelevant and suggested that physical forces were involved.

VII. CONCLUDING REMARKS

In this chapter we have discussed the development of mathematical models to describe the effect of temperature on the rate of bacterial growth. In Table 2 we show the inherent nonlinearity of Arrhenius plots and the model proposed should now replace the Arrhenius equation for modeling the temperature dependence of microbial growth rates.

In common with the Arrhenius equation, our models are empirical. We can, as yet, offer no physiological explanation for the square root response of bacterial growth rate to temperature. It is interesting to note that Equation 4 has recently been found to apply to the growth of plant cells.[40] This work relates the rate of plant growth to the fourth root of the nucleotide pool, while Koch[41] has shown that the specific growth rate of bacteria is related to the fourth root of the rate of ribosome synthesis. The physiological basis of the relationship awaits detailed studies in microbial physiology and genetics, particularly biosynthetic processes which are likely to be the key to the control of growth rate. The degree of uncoupling of catabolic and anabolic processes as temperatures diverge from the optimum also warrants consideration.

Despite the lack of a satisfactory physiological explanation, square root models are finding increased usage in microbiology. In Section V we outlined their use in monitoring psychrotrophic spoilage of foods. Similarly, they may be used to predict the growth rate of mesophilic and thermophilic organisms, e.g., in industrial fermentations.

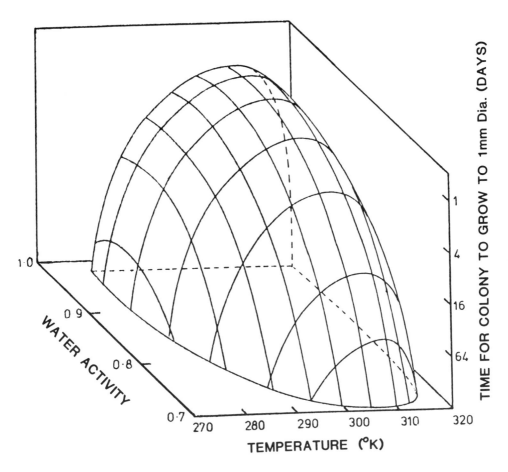

FIGURE 7. Time-temperature-water activity surface for *A. chevalieri* var. *intermedius*. (Doe, P. E., unpublished data.)

The finding that the square root equation described the effect of temperature on plant cell growth prompted a literature search on other complex biological processes. This revealed Belehradek's temperature function,[42-44] which is identical in form to the square root model. The latter is, in fact, a special case of Bělehrádek's function where the "temperature coefficient" is 2. Many examples of the relationship are cited by Belehradek[44] and in most cases the "temperature coefficient" lies between 1 and 3. The physiological basis of rate response to temperature was suggested to be governed by "protoplasmic viscosity".[43]

REFERENCES

1. **Ingraham, J. L., Maaløe, O., and Neidhardt, F. C.,** *Growth of the Bacterial Cell*, Sinauer Associates, Inc., Sunderland, Mass., 1983.
2. **Brock, T. D.,** *Thermophilic Microorganisms and Life at High Temperature*, Springer-Verlag, New York, 1978.
3. **Scott, W. J.,** Available water and microbial growth, in *Proceedings of Low Temperature Microbiology Symposium*, Campbell Soup Co., Camden, N.J., 1961, 89.
4. **Stetter, K. O.,** Ultrathin mycelia-forming organisms from submarine volcanic areas having an optimum growth temperature of 105°C, *Nature (London)*, 300, 258, 1982.

5. **Baross, J. A., Lilley, M. D., and Gordon, L. I.,** Is the CH_4, H_2 and CO_2 venting from submarine hydrothermal systems produced by thermophilic bacteria, *Nature (London)*, 298, 366, 1982.
6. **Baross, J. A. and Deming, J. W.,** Growth of black smoker bacteria at temperatures of at least 250°C, *Nature (London)*, 303, 423, 1983.
7. **Trent, J. D., Christian, R. A., and Yayanos, A. A.,** Possible artefactual basis for apparent bacterial growth at 250°C, *Nature (London)*, 307, 737, 1984.
8. **White, R. H.,** Hydrolytic stability of biomolecules at high temperatures and its implication for life at 250°C, *Nature (London)*, 310, 430, 1984.
9. **Campbell, A. M.,** Synchronization of cell division, *Bacteriol. Rev.*, 21, 263, 1957.
10. **Reichardt, W. and Morita, R. Y.,** Temperature characteristics of psychrotrophic and psychrophilic bacteria, *J. Gen. Microbiol.*, 128, 156, 1982.
11. **Mohr, P. W. and Krawiec, S.,** Temperature characteristics and Arrhenius plots for nominal psychrophiles, mesophiles and thermophiles, *J. Gen. Microbiol.*, 121, 311, 1980.
12. **Ratkowsky, D. A., Olley, J., McMeekin, T. A., and Ball, A.,** Relationship between temperature and growth rate of bacterial cultures, *J. Bacteriol.*, 149, 1, 1982.
13. **Johnson, F. H., Eyring, H., and Stover, B. J.,** *The Theory of Rate Processes in Biology and Medicine*, John Wiley & Sons, New York, 1974, 199.
14. **Ratkowsky, D. A., Lowry, R. K., McMeekin, T. A., Stokes, A. N., and Chandler, R. E.,** Model for bacterial culture growth rate throughout the entire biokinetic temperature range, *J. Bacteriol.*, 154, 1222, 1983.
15. **Olley, J. and Ratkowsky, D. A.,** Temperature function integration and its importance in the storage and distribution of flesh foods above the freezing point, *Food Technol. Aust.*, 25, 66, 1973.
16. **Olley, J. and Ratkowsky, D. A.,** The role of temperature function integration in monitoring of fish spoilage, *Food Technol. N.Z.*, 8, 13, 1973.
17. **Ohta, F. and Hirahara, T.,** Rate of degradation of nucleotides in cool-stored carp muscle, *Mem. Fac. Fish. Kagoshima Univ.*, 26, 97, 1977.
18. **Shaw, M. K., Marr, A. G., and Ingraham, J. L.,** Determination of the minimal temperature for growth of *Escherichia coli*, *J. Bacteriol.*, 105, 683, 1971.
19. **McMeekin, T. A.,** Preliminary observations on heterotrophic, psychrotrophic and psychrophilic bacteria isolated from Antarctic water samples, submitted for publication in *Biological Research in the Vestfold Hills*, Burton, H., Johnson, G., and Bayly, I., Eds., Dr. W. Junk, The Hague, in press.
20. **Chandler, R. E. and McMeekin, T. A.,** Temperature function integration and its relationship to the spoilage of pasteurised, homogenised milk, *Aust. J. Dairy Technol.*, 40, 10, 1985.
21. **Smith, M. G.,** Generation time, lag time and minimum growth temperature of growth of coliform organisms on meat, and the implications for codes of practice in abbatoirs, *J. Hyg. Cambridge*, 94, 289, 1985.
22. **Marr, A. G., Ingraham, J. L., and Squires, C. L.,** Effect of the temperature of growth of *Escherichia coli* on the formation of beta-galactosidase, *J. Bacteriol.*, 87, 356, 1964.
23. **Shaw, M. K.,** Effect of abrupt temperature shift on the growth of mesophilic and psychrophilic yeasts, *J. Bacteriol.*, 93, 1332, 1967.
24. **Herendeen, S. L., Van Bogelen, R., and Neidhardt, F. C.,** Levels of major proteins of *Escherichia coli* during growth at different temperatures, *J. Bacteriol.*, 139, 185, 1979.
25. **Pooni, G. S. and Mead, G. C.,** Prospective use of temperature function integration for predicting the shelf life of non-frozen poultry meat products, *Food Microbiol.*, 1, 67, 1984.
26. **Owen, D. and Nesbitt, M.,** A versatile time-temperature function integrator, *Lab. Pract.*, 33, 70, 1984.
27. **Gill, C. O.,** Prevention of early spoilage of livers, in *Proc. Eur. Mtg. Meat Research Workers Conf.*, Bristol, No. 30(5), 14, 1985.
28. **Olley, J. and McMeekin, T. A.,** Prediction of histamine formation based on time temperature history, in *Histamine in Marine Food Products: Production by Bacteria, Measurement and Prediction of Formation*, Bonnie Sun Pan and James, D. G., Eds., Food and Agriculture Organization, Rome, 1985, 252.
29. **Olley, J.,** Temperature effects on histamine producing bacteria, in *Properties and Processing of Marine Foods*, Marine Food Science Series, Bonnie Sun Pan, Ed., National Taiwan College of Marine Science and Technology, Keelung, 1983, 14.
30. **Chandler, R. E. and McMeekin, T. A.,** Temperature function integration and the prediction of the shelf life of milk, *Aust. J. Dairy Technol.*, 40, 37, 1985.
31. **Gibson, D. M.,** Predicting the shelf life of packaged fish from conductance measurements, *J. Appl. Bacteriol.*, 58, 465, 1985.
32. **Spencer, R. and Baines, C. R.,** unpublished data, 1964, reappraised by R. M. Storey and D. Owen, 1982 (with permission of the latter authors).
33. **Gibbs, P. A., Williams, A., and Stannard, K.,** personal communication.
34. **Doe, P. E.,** Spoilage of dried fish — the need for more data on water activity and temperature effects on spoilage organisms, in *Proceedings of the Workshop on the Production and Storage of Dried Fish*, FAO Fish Rep. No. 279, Suppl., James, D., Ed., Food and Agriculture Organization, Rome, 1983, 209.

35. **Ayerst, G.,** The effects of moisture and temperature on growth and spore germination in some fungi, *J. Stored Prod. Res.,* 5, 127, 1969.

36. **Olley, J., Doe, P. E., and Heruwati, E. S.,** The influence of drying and smoking on the nutritional properties of fish; an introductory overview, in *The Effect of Smoking and Drying on the Nutritional Properties of Fish,* Burt, J. R., Ed., Min. Agric. Fish and Food, Torry Res. Stn. Aberdeen, Scotland, 1987, chap. 1.

37. **Ohye, D. F. and Christian, J. H. B.,** Combined effects of temperature, pH and water activity on growth and toxin production by *Clostridium botulinum* types A, B and E, Botulism 1966, in *Proc. 5th Int. Symp. Food Microbiol.,* 1967, 217.

38. **Thomas, C.,** personal communication, 1984.

39. **Gill, C. O. and Penney, N.,** Penetration of bacteria into meat, *Appl. Environ. Microbiol.,* 33, 1284, 1977.

40. **Mingo, R. and Lopez-Saez, J. F.,** Relationships between growth temperature, ATP level and proliferation rate of higher plant cells, personal communication.

41. **Koch, A. L.,** Overall controls on the biosynthesis of ribosomes in growing bacteria, *J. Theor. Biol.,* 28, 203, 1970.

42. **Bělehrádek, J.,** Influence of temperature on biological processes, *Nature (London),* 118, 117, 1926.

43. **Bělehrádek, J.,** Protoplasmic viscosity as determined by a temperature coefficient of biological reactions, *Nature (London),* 118, 478, 1926.

44. **Bělehrádek, J.,** Temperature and Living Matter, Protoplasma Monogr., No. 8, Gebruder Borntraeger, Berlin, 1935.

Chapter 5

POPULATION GROWTH KINETICS OF PHOTOSYNTHETIC MICROORGANISMS

Yuan-Kun Lee

TABLE OF CONTENTS

I. INTRODUCTION

Being primary producers, photosynthetic microorganisms play an important role in maintaining the energy and carbon balance in natural ecosystems. Their relatively simple structure and rapid rates of growth render them favorable subjects in the study of photosynthetic reactions.[1] In the past 35 years, interest in photosynthetic microorganisms has extended to their mass cultivation as a means of harnessing solar energy,[2,3] producing food, feed and chemicals,[4,5] and sewage treatment.[6] A better understanding of the population kinetics of photosynthetic microorganisms in culture is crucial in the elucidation of their photosynthetic mechanisms, in environmental production studies, and also in the design of effective mass culture systems. Growth models of nutrient-limited microalgae and photosynthetic bacteria have been reviewed extensively by Rhee,[7] Rhee et al.,[8] and Cunningham and Maas.[9] In this chapter, the concentration is on the effect of incident light on the population growth kinetics of photosynthetic microorganisms.

II. GROWTH IN FED BATCH CULTURE

The unique characteristic of photosynthetic cultures is that the energy source for growth is not a constituent of the culture medium, but is supplied as the flux of quantum energy at the surface of the culture. A portion of the light energy that impinges on a photosynthetic culture system is absorbed by the culture with the rest either reflected or passing directly through it. Growth of the culture depends upon the continuous supply of light, and the biomass output rate is a function of the rate of energy supply. Thus, strictly speaking, there is no batch photosynthetic system, instead a constant volume fed batch system.

A. Light Sufficient Growth

In a fed batch culture, under continuous illumination at constant intensity and with all nutrients in excess, the population will grow at its maximum specific growth rate (μ_m), after an initial lag phase, as long as the photosynthetically available radiance (PAR) exceeds the maximum light energy required for growth, i.e., $\phi I_o > \mu_m XV/YA$, where ϕ = fraction of incident intensity available for photosynthesis, I_o = incident light intensity, X = biomass density, V = total culture volume, Y = growth yield on light, and A = illuminated culture surface area. (See B in Figure 1).

B. Saturation Light Intensity

The minimum light intensity required to attain the maximum specific growth rate of a photosynthetic culture is called the saturation light intensity (I_K). Attempts have been made to measure the I_K value of photosynthetic cultures from plots of specific growth rate against incident light intensity.[10-14] For a variety of freshwater and marine algae I_K has been reported to range from 13.9 to 4.18 Wm^{-2}. However, by virtue of its definition, $I_K = \mu_m XV/YA$, it appears more relevant to estimate I_K from the energy balance for a photosynthetic culture, since I_K is a function of X, V, Y, and A.

Let us first consider the energy balance equation of a single spherical photosynthetic cell. If the photosynthetic cell has a peripheral light harvesting apparatus with outer radius equivalent to the radius of the cell, r, its surface area will be $4\pi r^2$. If the cell is illuminated from one side, the effective area for light reception will be one half of the total area. At the point of light saturation, the light input equals the light absorbed, and the energy balance is $I_K 2\pi r^2 = \mu M/Y$. Hence, the saturation light intensity is

$$I_K = \mu M/2\pi r^2 Y = 2\mu r\rho/3Y \tag{1}$$

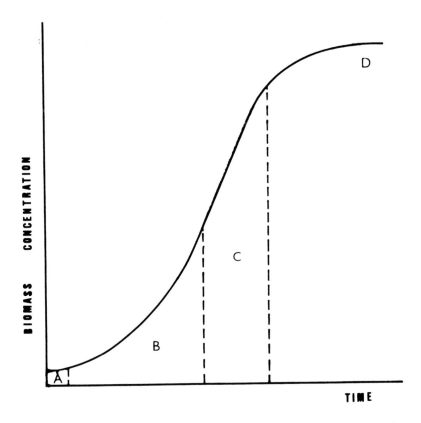

FIGURE 1. Fed batch growth curve of a photosynthetic culture with four phases: A, lag; B, exponential growth; C, linear growth; D, stationary.

where M = mass of a cell and ρ = the cell mass density. For *Chlorella vulgaris*, the different parameters in Equation 1 have been reported.[3] The maximum value of μ was found to be 6.1×10^{-5}/sec, the average r at μ_m was 2×10^{-6} m, ρ was 1.1×10^3 kg/m^3, and BY was estimated at 1.53×10^{-8} kg/J. Substituting these values into Equation 1 gives I_K = 5.85 Wm^{-2} as the required minimum photosynthetically available radiance (PAR) to saturate the algal cell. In this way it is possible to calculate the I_K value of *Chlorella* cultures at different cell concentrations. These are represented in Figure 2. It can be observed that the I_K value derived in this way is different for different biomass densities. The I_K of a photosynthetic culture is a useful physiological parameter. For instance, an increase in I_K over the theoretical value estimated from Equation 1 would indicate either a reduction in Y or as an increase in rρ. The latter would imply an alteration in the cellular composition.

C. Light-Limited Growth

A photosynthetic culture will continue to grow exponentially until all the photosynthetically available radiation is used for growth, i.e., $\phi I_o = \mu_m XV/YA$. Thereafter the biomass accumulates at a constant rate, μX, until some substrate in the culture medium becomes the limiting factor (portion C and D of Figure 1). The time when the exponential phase ceases and linear growth begins depends on the incident light intensity, the size of inoculum, and the viability of the culture. The higher the light intensity and the lower the inoculum density, the longer the exponential phase. The linear growth rate is largely dependent on the light intensity.

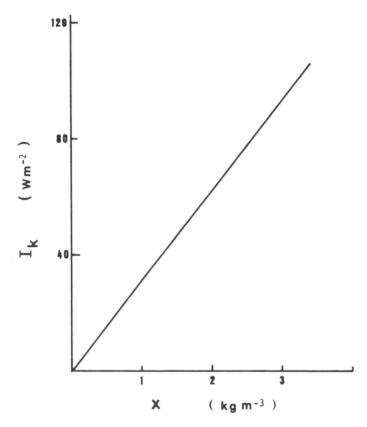

FIGURE 2.   Saturation light intensity (I_K) of photosynthetic *Chlorella* cultures at various biomass concentrations (X) according to the relation $I_k = \mu_m XV/AY$, $\mu_m = 6.1 \times 10^{-5}$/sec, $V/A = 7.85 \times 10^{-3}$m, $r = 2 \times 10^{-6}$ m, $Y = 1.53 \times 10^{-8}$ kg/J.

III. GROWTH IN CHEMOSTAT CULTURE

A. Steady-State Kinetics

The chemostat has been widely used in the study of photosynthetic microorganisms.[3,7-9,15,16] A chemostat is a constant volume culture system where fresh medium is fed and harvested continuously at the same rate. The principles of chemostat[17] indicate that a self-regulating steady state in which the biomass concentration remains constant will eventually be reached. At steady state, the dilution rate (D = F/V where F = medium flow rate) is equal to the specific growth rate of the culture. An ideal photosynthetic chemostat is one with homogeneously distributed light intensity in the culture system. However, due to the way in which light penetrates a suspension, light energy is distributed in reducing amounts toward the center of the culture. In practice, near ideal chemostat conditions can be achieved either by having a thin culture of low cell density so that most of the light energy can pass through the culture, or having a high degree of mixing so that the cells move rapidly across the intensity gradient.

In a light-limited photosynthetic chemostat, assuming all incident PAR is absorbed by suspended cells, then:

The net increase in energy content of the culture	=	The energy absorbed by biomass	−	The energy in out-flow biomass

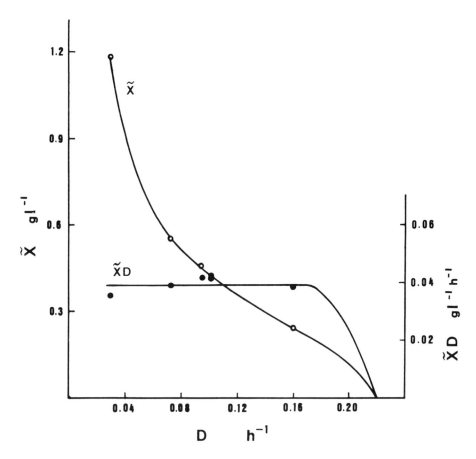

FIGURE 3. Steady-state biomass concentration ($\widetilde{X}$ and biomass output rate ($\widetilde{X}D$) as functions of the dilution rate (D) in chemostat cultures of *Chlorella vulgaris* at constant light intensity of 2.91 kJ/D/hr. μ_m was determined by washout method.[3]

For an infinitely small time interval, dt, the balance for the whole culture is

$$VdE = \phi I_o Adt - FXdt/Y \quad \text{or} \quad dE/dt = \phi I_o A/V - FX/YV$$

where dE = increases in energy content of biomass.

In the steady state, dE/dt = 0, and D = F/V. Hence,

$$D\widetilde{X} = \phi I_o AY/V \tag{2}$$

where the tilde denotes a steady state value. Pirt and co-workers[3] working on a light-limited *Chlorella vulgaris* culture, were able to confirm this relationship. The plots of steady state biomass concentration ($\widetilde{X}$) and biomass output rate ($D\widetilde{X}$) of a *Chlorella* culture are shown in Figure 3. It is clear from the plots that for a given light intensity, $D\widetilde{X}$ is virtually constant, implying that Y is also constant.

The relation between biomass output rate and the volumetric energy input for a *Chlorella* culture is presented in Figure 4. The mean biomass output rate increased linearly with volumetric energy input, and the plot appeared to pass through the origin, indicating that the saturation constant of the light was very small.

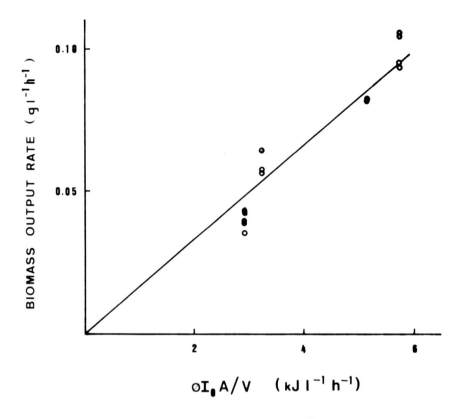

FIGURE 4.   Steady-state biomass output rate of *Chlorella vulgaris* ($\widetilde{X}$D) as a function of incident light energy per unit volume ($\emptyset I_oA/V$) in cheomstat culture.[3]

B. Light Saturation Constant

The relationship between the light energy absorbed (I) and the specific growth rate of a photosynthetic culture has the form of the Monod relation,[14]

$$\mu = \mu_m I/(I + K_I) \tag{3}$$

where K_I = light saturation constant. Rearranging the above equation, one obtains

$$1/\mu = K_I/\mu_m I + 1/\mu_m \tag{4}$$

By measuring the steady-state mean irradiance absorbed by a culture of photosynthetic bacterium, *Rhodopseudomonas capsulata*, and its specific growth rate Göbel[14] was able to estimate the K_I values for light at wavelengths of 860 and 522 nm.

Under the conditions of the experiments, the results of which are shown in Figure 5, it was found that for monochromatic light at 860 nm, where irradiance is mainly absorbed by bacteriochlorophyll, the K_I was 2.5 nEinstein/sec/cm^2. The K_I value of accessory carotenoid pigments (522 nm) was 10.3 nEinstein/sec/cm^2. Apparently bacteriochlorophyll has a higher affinity for light than carotenoid. The μ_m of the *Rhodopseudomonas* culture was found to be 0.33/hr and independent of the wavelength used.

It is worth mentioning that in the same report,[14] both bacteriochlorophyll and carotenoid contents were found to be functions of mean irradiation, i.e.,

$$P = P_{min} + K_I(P_{max} - P_{min})/(K_I + I_o) \tag{5}$$

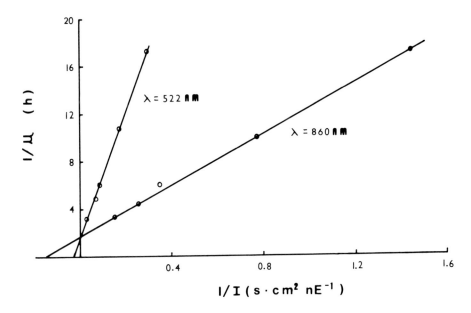

FIGURE 5.   Reciprocal of specific growth rate (μ) of *Rhodopseudomonas capsulata* culture plotted as a function of irradiance absorbed (I) at wavelengths 860 and 522 nm according to the relation $1/\mu = K_I/\mu_m\, 1 + 1/\mu_m$. Mineral medium with lacate as the carbon source, temperature 30°C.[14]

where P_{max} is the maximum pigment content, and P_{min} the minimum pigment content. At light saturation, where $I_o \gg k_1$, $P = P_{min}$. In the dark, $P = P_{max}$ since now $I_o = 0$. It appears that the light harvesting pigment content of a photosynthetic culture does not affect the affinity (K_I) and rate of uptake of light energy. The alteration in the steady-state pigment content at different dilution rates is a reflection of the physiological adaptation of the culture to absorb all available PAR.

Among the algae, the green algae seem to have a lower affinity for white light compared to other groups, the K_I values decreasing in the order of chlorophytes > diatoms > dinoflagellates. Such a relationship has been used to suggest the significance of the K_I values in determining the distribution of algae in natural ecosystems.[18]

IV. DEVIATION FROM STANDARD GROWTH KINETICS

A. Effect of Maintenance Energy Requirement

Like all microbial cells, photosynthetic microorganisms require energy for growth and for purposes other than growth. Some maintenance functions which have been identified are turnover of cellular components, osmotic work to maintain solute gradients across membranes, and cell motility.[19] When the fraction of energy consumed for nongrowth functions is significant compared to the total energy consumption, the apparent growth yield, Y, becomes a function of specific growth rate,[3,19] i.e.,

$$1/Y = 1/Y_G + m/\mu \quad \text{or} \quad Y = Y_G\mu/(\mu + mY_G) \tag{6}$$

where Y_G = maximum growth yield on light, and m = maintenance energy coefficient. Hence,

$$\mu = \phi I_o A Y_G/VX - mY_G \tag{7}$$

$$\mu X = \phi I_o A Y_G/V - mXY_G \tag{8}$$

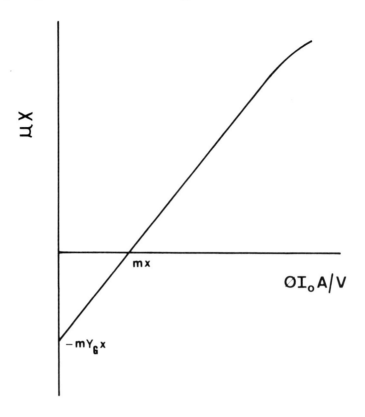

FIGURE 6.   Relation between biomass output rate (μX) and incident light energy ($\emptyset I_o A/V$) (Equation 8) assuming a constant maintenance coefficient (m). Note that when $\mu X = 0$, $\emptyset_o A/V = mX$, and when $\emptyset I_o A/V = 0$, $\mu X = -mY_G X$.

The maximum specific growth is decreased by mY_G if there is a requirement for maintenance energy (Equation 7). The graphical representation of Equation 8 is given in Figure 6.

1. Fed Batch Culture

It can be seen from Equation 8 that the effect of maintenance energy requirement is to decrease the gradient of the growth curve in a fed batch culture by mXY_G (Figure 7).

2. Chemostat Culture

In the chemostat, the requirement for maintenance energy by photoautotrophs can be expressed by substituting $Y = Y_G \mu/\mu + mY_G$ in the energy balance equation, i.e., $dE/dt = \emptyset I_o A/V - DX/Y_G - mX$. In steady state,

$$\phi I_o A/V = D\tilde{X}/Y_G = m\tilde{X} \tag{9}$$

or

$$\tilde{X} = \phi I_o A Y_G/V(D + mY_G) \tag{10}$$

The effect of maintenance energy requirement on the steady state biomass concentration and biomass output rate is shown in Figure 8. The deviations of both the biomass density ($\tilde{X}$) and the biomass output rate ($\tilde{X}D$) from the ideal condition are greatest at lower dilution

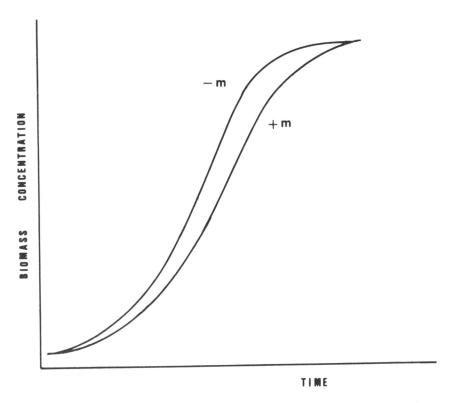

FIGURE 7.   Growth curves of fed batch culture of a photosynthetic culture when there is negligible maintenance energy requirement (− m) and when there is a significant maintenance energy requirement (+ m) for the growth-limiting substrate.

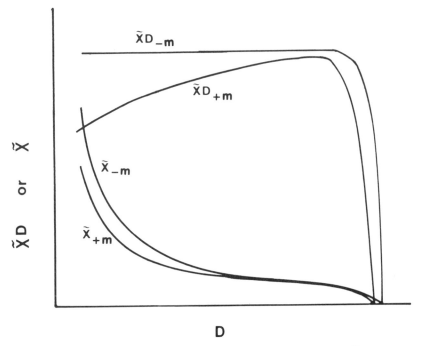

FIGURE 8.   Steady-state biomass output rate ($\widetilde{X}D$) and biomass concentration ($\widetilde{X}$) as functions of dilution rate (D) in a chemostat. − m denotes negligible maintenance energy requirement and + m a significant energy requirement for the growth-limiting substrate.

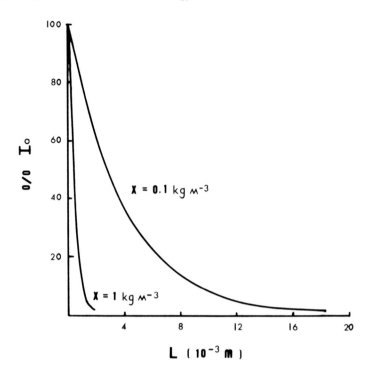

FIGURE 9.   The penetration of light of wavelength 680 nm into a *Chlorella* culture; a $= 1.09 \times 10^{-3}$ m³/kg.

rates. This is attributed to the fact that at lower D values, the maintenance energy constitutes a larger fraction of the overall energy requirement.

B. Effect of Short Intermittent Illumination

If a beam of monochromatic radiation of intensity I_o photons per unit area per second is incident to a photosynthetic cell suspension of density X, and the intensity transmitted is I, the relationship of absorbance and cell density can be described by the Beer-Lambert Lawof absorption, i.e., $\log_{10}(I_o/I) = aXL$, where a = absorptivity of the sample at the wavelength considered and L = light path.

For a culture suspension at a concentration 0.1 kg/m³ with biomass absorptivity of 1.09 $\times$ 10³ m²/kg at 680 nm wavelength, 99% of the light can penetrate 4 $\times$ 10⁻⁵ m (0.04 cm) into the culture next to the illuminated surface (Figure 9), whereas for a culture of 1 kg/ m³, 99% of the incident light at 680 nm will penetrate only 4 $\times$ 10⁻⁶ m into the culture. These simple calculations show that under normal culture conditions where the thickness or the diameter of the culture vessel is many centimeters, it is usually unavoidable for part of the photosynthetic culture to be under light-limited conditions at any one time. As a consequence, cells circulating in the vessel receive an intermittent supply of energy. A dark fraction in a culture system can also result from the design of the bioreactor, such as when a degassing chamber to accomodate sensor probes in a loop reactor[20] or in a channel type reactor system incorporates a sump and tower.[21]

The effect of short intermittent illumination on the energetics of *Chlorella* cultures was studied by Lee and Pirt[22] in a tubular loop reactor, where part of the loop was blacked out to create a dark zone. A short intermittent illumination was defined as one which is not long enough to allow a substantial change in cellular composition during the cycle.

1. Light-Limited Culture

Under light-limited conditions, it was observed experimentally[22] that the apparent main-

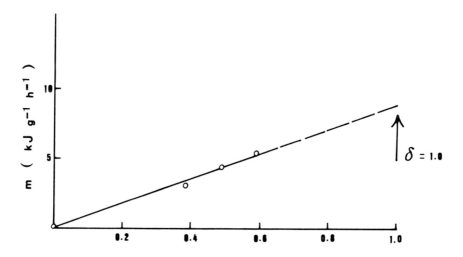

FIGURE 10.   Maintenance energy coefficient (m) as a function of the dark fraction (δ) in light-limited fed batch cultures of *Chlorella vulgaris*[22]

tenance energy coefficient (m) of an algal culture was a function of the dark fraction (δ) (Figure 10),

$$m = \delta m_r \tag{11}$$

where m_r = maintenance coefficient of resting cells. Hence, the energy balance for photosynthesis becomes

$$\phi I_o = \mu_o XV/AY_G + m_r XV/A,$$

where μ_o is the overall (apparent) specific growth rate of the algal culture. Rearrangement of the above equation gives

$$\phi I_o A/XV = q = \mu_o/y_G + \delta m_r \tag{12}$$

where q = the specific rate of light utilization by the biomass.

The actual growth yield (Y) from the light utilized can be deduced from Equation 12. Since $q = \mu_o/Y$ and $mY_G = \mu_e$, it follows that

$$Y = Y_g \mu_o/(\mu_o + \mu_e) \tag{13}$$

Thus, the greater the dark fraction (δ) and the specific maintenance rate (μ_e), the smaller the actual yield. When $\mu_o = \mu_e$, $Y = 0.5 Y_G$. The specific growth rate in the photostage during light-limited growth is obtained from the biomass balance, $\mu_o XV = \mu X \alpha V - \mu_e X \delta V$, where light fraction $\alpha = 1 - \delta$, hence,

$$\mu = (\mu_o + \delta\mu_e)/\alpha \tag{14}$$

In a light-limited chemostat, the steady-stage biomass concentration is given by

$$\tilde{X} = \phi I_o AY_G/V(D + \delta\mu_e) \tag{15}$$

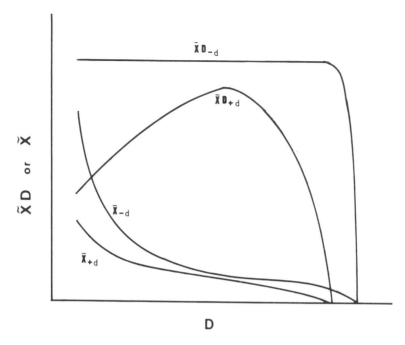

FIGURE 11. Steady-state biomass output rate ($\widetilde{X}D$) and biomass concentration ($\widetilde{X}$) of a photosynthetic culture in a cheomstat with ($+d$) and without ($-d$) a dark fraction.

Equation 15 indicates that the steady-state biomass is inversely proportional to the dark fraction, which in turn is a function of the biomass concentration. The effect of a dark fraction in a light-limited culture is then more prominent at lower dilution rates, where the biomass concentration is higher (Figure 11). With increasing dilution rate, a situation will be reached where D becomes greater than μ_o, resulting in the wash-out of the cells. This wash-out is, however, transient since the decrease that accompanies the wash-out results in an increasing μ_o (Equations 14 and 15). Thus, when μ_o becomes equal to D, a new steady state with a lower biomass density will be established (Figure 11). This parabolic relationship between $D\widetilde{X}$ and the dilution rate in a photosynthetic chemostat, with a peak productivity at a growth rate somewhat lower than μ_m had been repeatedly reported with laboratory cultures[23—27] and outdoor cultures.[12,13,28] It is worth nothing, however, that a decreased steady-state biomass output rate at high dilution rate in a photosynthetic chemostat culture could also be the result of trace element[29] or gaseous substrate limitation.

The detrimental effect of a dark fraction in a light-limited photosynthetic culture could be partially removed by increasing the degree of mixing the culture system, i.e., to reduce the duration of the intermittent cycle.[14,28] On the other hand, it had been reported that at high incident light intensity, a very short intermittent cycle (of the order of milliseconds) allowed the full efficient utilization of all the energy absorbed in the short photostage, which intensity otherwise proved saturating or even inhibiting.[26,30]

2. Light Sufficient Exponential Growth

A dark stage could also be incurred in a light sufficient culture system. When light is sufficient, the cells in the photostage will grow at the maximum specific growth rate. Upon entering the dark stage, the cells may continue to grow on the reserve energy stored during the photostage. The biomass balance equation then becomes:

| Net increase in biomass | = | Growth in light zone | + | Growth in dark zone | − | Loss due to endogenous metabolism in the dark rest zone |

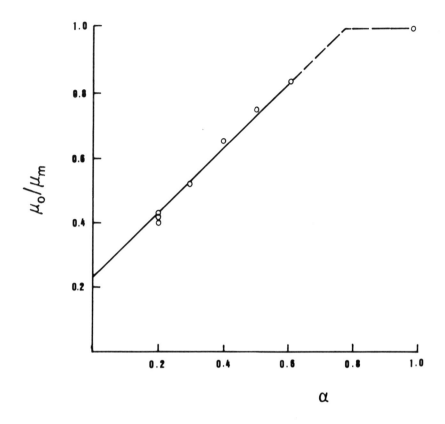

FIGURE 12.   Plots of the ratio of the observed growth rate to the maximum specific growth rate (μ_o/μ_m) as a function of the light fraction (α) for *Chlorella vulgaris* in fed batch culture.[22]

or $\mu_o\,XV = \mu_m X\alpha V + \mu_m X\alpha V - \mu_e X(1-(\alpha+\beta))V$, where α = fraction of culture in the illuminated zone, β = fraction which fixed CO_2 in the dark (growth in the dark), δ = fraction of culture in the dark. It is assumed that the specific maintenance rate (μ_e) causes degradation of the biomass in the dark zone only. Rearrangement of the last equation gives

$$\mu_o = (\mu_m + \mu_e)\,\alpha + (\mu_m + \mu_e)\,\beta - \mu_e \qquad (16)$$

Hence, the first indication of the effect of a dark fraction in a culture system is a reduction in the maximum specific growth rate of the culture.[31] From Equation 16, a plot of μ_o against α should give a straight line with slope $(\mu_m + \mu_e)$ from which μ_e can be obtained. Also, it follows that when $\mu_o = \mu_m$, $\beta = 1 - \alpha$. If $\mu_e = 0$, then the equation can be rearranged to become

$$\mu_o/\mu_m = \alpha + \beta \qquad (17)$$

and a plot of μ_o/μ_m against α should have an intercept β at $\alpha = 0$. The predictions of Equations 16 and 17 were tested experimentally and verified by Lee and Pirt[22] (Figure 12). With the green algal culture (MA003) used in their work, it was shown that growth of a light sufficient culture continued in the dark for 3.6 sec at the maximum rate.

C. Effect of Culture Temperature

It was observed in *Chlorella* and *Chlorogonium*[32] cultures that the maximum growth yield

(Y_G) was a function of temperature, increasing with temperature up to the optimal temperature for growth. This relationship can be represented empirically by

$$Y_G = de^{bT} \tag{18}$$

where Y_G has the units gJ^{-1}, T is the culture temperature in degrees centigrade, while b and d are constants (found to be $0.0773°C^{-1}$, $9.85\ gJ^{-1}$ $0.0677°C^{-1}$, and $7.36\ gJ^{-1}$ respectively for *Chlorella vulgaris* and *Chlorogonium* sp.).

In contrast, m appears to be inversely proportional to the culture temperature, and can be represented empirically by the relation

$$m = ge^{-fT} \tag{19}$$

where m is in gJ^{-1}/hr. The values of g and f for *Chlorella vulgaris* and *Chlorogonium* sp. were estimated to be $6.90\ gJ^{-1}/hr$, $0.166°C^{-1}$ and $2.96\ gJ^{-1}/hr$, and $0.052°C^{-1}$, respectively. Incorporating Equations 18 and 19 into Equation 9, the energy balance becomes

$$\phi I_o A/V = \mu X/de^{bT} + Xge^{-fT} \tag{20}$$

It is not clear whether these effects of temperature on Y_G and m are common to other photosynthetic microorganisms. As both Y_G and m are related to the specific maintenance rate by the equation $\mu_e = mY_G$, a number of workers[33] have measured the specific maintenance rate (μ_e) of microalgae at various temperatures and found it to be relatively insensitive with only a slight upward trend toward the higher temperature range.[33] The same situation, however, may not be expected of both m and Y_G, since a constant μ_e with respect to temperature could be the result of opposing effects by temperature on m and Y_G. This appeared to be so in the earlier example.

The enhancing effect of culture temperature on the biomass yield of a light-limited algal culture has been attributed to two causes.[32] First, at low temperatures, there was a change in the composition of the biomass with more energy being expended to synthesize a higher protein content. Second, at low temperatures, a cyanide-resistant respiratory pathway becomes operative which led to a decrease in the number of ATP molecules being generated. The changes in respiratory pathways, however, were not reflected in the activation energy, as the effect of culture temperature on μ_m could be described by an Arrhenius type rate-temperature model up to the optimal temperature for growth.[32] The inverse relationship between μ_m and growth temperature might reflect energy wastage by the algal cells at low temperature.[32]

D. Effect of Culture pH

At constant incident light intensity, when the pH of a cyanobacterium (*Oscillatoria agardhii*) culture was increased from 8.0 to 9.0,[16] it was observed that both the photosynthetic efficiency (PE = Y_GK, where K is the energy content of biomass) and specific maintenance rate ($\mu_e = mY_G$) increased. Therefore, it is envisaged that in a fed batch culture without pH control, the slope of the growth curve will depend on the pH of the culture due to the consumption of substrate or release of algal products.

IV. GROWTH IN CYCLOSTAT CULTURE (DIURNAL INTERMITTENT ILLUMINATION)

In natural environments, photosynthetic microorganisms are subjected to diurnal illumination of varying intensity from dawn to dusk and then (almost) complete darkness. The

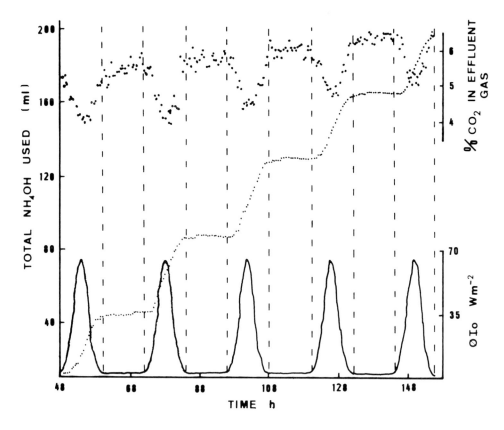

FIGURE 13.   The effect of diurnal intermittent illumination of varying intensity ($\emptyset I_o$) on CO_2 and NH_4OH uptake of a *Chlorella* culture.[39]

growth kinetics of microalgae under such conditions have been studied mostly in the laboratory in alternate periods of constant light intensity and darkness. In these studies, cyclic changes in cell number, cell volume, dry weight, storage energy, and carbohydrate content have been observed in the light-dark cycle.[8,16,33—35] In a photosynthetic cyclostat, which is a chemostat with diurnal intermittent illumination, the overall relationship between D and $\widetilde{X}$ can be described by Equation 9, if $\widetilde{X}$ is taken to be the average biomass over the period of the light-dark cycle.[8,16,33] Hence, D determines only the integrated instantaneous growth rate average over the period P, and $D_p = \int_o^P \mu(t)dt$, where D_p is the integrated specific growth rate over period P. Whereas energy is provided by light in the light cycle, cells in the dark cycle must rely on endogenous metabolism for maintenance purpose. Gons and Mur[33] reported that the decrease in biomass of *Scenedesmus protuberans* in the dark cycle was in the order of the specific maintenance rate. Other workers[36-38] observed in light-limited cyclostat cultures of *Thalassiosira fluviatilis* and *Skeletonema costatum* that the rate of biomass carbon loss in the dark (C) due to endogenous metabolism was a linear function of the specific growth rate of the cultures, $C = i + h\mu$, where i and h are constants. For the *Thalassiosira* culture, the values of i and h were 0.046/day and 0.198, whereas for *Skeletonema*, i and h were 0.068/day and 0.47.

To simulate the diurnal variation in light intensity, Pirt and co-workers[20,39,40] developed a system where computer-controlled motorized Venetian blinds were placed between the light source and a tubular loop bioreactor. The growth of a *Chlorella* culture in the bioreactor was monitored continuously (Figure 13). Detailed analysis of one of the light cycles (118.5

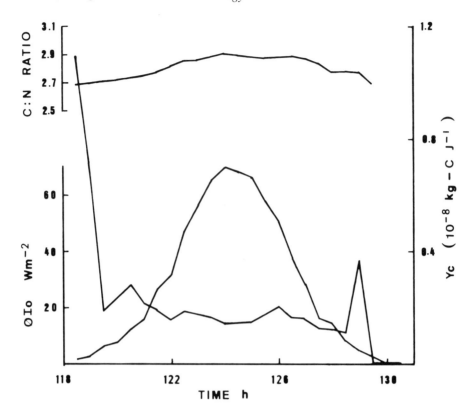

FIGURE 14.   The C:N ratio of the *Chlorella* biomass and carbon yield on light energy absorbed (Y$_c$) calculated from one of the light cycles (118.5 to 130.5 hr) in Figure 13.

to 130.5 hr, Figure 13) reveals that the C:N ratio of the *Chlorella* biomass varied between 2.68 at the onset of the light cycle and 2.91 when the light intensity was at the highest (Figure 14). These values are, however, much lower than those (C:N = 6) in the steady-state biomass of the *Chlorella* cells growing in continuously illuminated chemostat.[3] It appears that under a diurnally intermittent illumination regimen, the algal cells actively accumulate N-substrate. Ammonia uptake occurred only in the light cycle, and therefore was energy dependent. However, it was independent of the specific growth rate.

The uptake of CO_2 by the algal culture in the light cycle coincided with the pattern of light intensity implying light-limited growth (Figure 13). When C-yield on incident light was calculated from the data reported[39] a rather unexpected result was found. At the onset of the light cycle, the C-yield value was at the highest (Figure 14), and equal to the value obtained from continuous illumination experiments.[3] The yield dropped drastically in the following hour, and continued to decrease at a slower rate until the end of the light cycle. There are three conceivable explanations for these observations:

1.   The use of Venetian blinds to vary the average light intensity that arrives at the surface of the culture reactor created a dark fraction in the culture system and this resulted in algal cells receiving intermittent illumination while traveling along the tubular reactor. The reduced overall growth yield would therefore be an expression of the decreased apparent specific growth rate (μ_o, Equation 13) and increased maintenance energy coefficient (m, Equation 10). This explanation could be confirmed by increasing the culture flow rate in the tubular reactor, so that the average mean time a cell spends in the dark fraction would be reduced.

2. The rate-limiting step in the utilization of light of varying intensity could be the readjustment of the light harnessing mechanism, i.e., the photoelectric excitation and photochemical reactions, in response to the changing light intensity. If this was the case, then by reducing the rate of change in intensity in the light cycle, a state could be reached when the culture was able to cope with the varying intensity and utilize all energy absorbed.

3. The low C:N ratio of the algal biomass implies active intracellular conservation of ammonia. Ammonia at high concentration is known to be toxic to microbial cells. To remove the inhibitory effect of excess intracellular ammonia, its supply could be limited to specific periods in the light cycle so as to encourage the utilization of the conserved N-substrate.

More fundamental research needs to be done before the growth kinetics and physiology of photosynthetic cells growing in diurnal intermittent illumination can be fully understood. Only then can meaningful kinetic models be constructed.

VI. PHOTOINHIBITION AT HIGH LIGHT INTENSITY

The depression of the maximum specific growth rate and the impairment of photosynthetic capabilities of photosynthetic algae and bacteria by high intensity visible light has been widely documented.[41-45] A number of photodynamic effects have been proposed.[42,46-50] Results seem to indicate that photoinhibition occurs when the potentials of various biophysical and biochemical systems for dissipating excitation energy are exceeded, and an excess of electrons in the photosynthetic reaction centers results. Assuming that photoinhibition results from the accumulation of an intermediate of energy metabolism, the photoinhibition effect may be described in the form of the equation used for noncompetitive inhibition by a product.[18]

$$\mu = \mu_m \phi I_o / \gamma (K_I + \phi I_o) \tag{20a}$$

$$\gamma = 1 + q_e/K_i \tag{21}$$

where q_e is the specific rate of accumulation of the energy intermediate and K_i is the specific inhibition constant. Both q_e and K_i have the unit of energy per biomass per time, e.g., J/kg/sec. Also,

$$q_e = Y_{e/l} \phi I_o A/VX - Y_{e/x}\mu - Z \tag{22}$$

The first term on the right hand side of the equation represents the specific rate of production of the energy intermediate, where $Y_{e/l}$ is the production yield of the intermediate on light energy absorbed. The middle term represents the specific rate of conversion of this energy intermediate to biomass. Here, $Y_{e/x}$ is the growth yield of biomass on the energy intermediate and Z represents the highest possible capacity of the photosynthetic cells to dissipate the energy intermediate by biophysical and biochemical means not related to growth. It has been suggested and generally accepted that the major function of carotenoids in photosynthetic cells is to protect cells against potentially harmful or lethal photodynamic effects through their ability to quench excess light excited sensitizers.[42] Hence, Z in Equation 22 can be determined largely by the carotenoid content of the photosynthetic culture. At the intensity when light becomes inhibiting, i.e., $\phi I_o \gg K_I$,

$$\mu = \mu_m/\gamma \tag{23}$$

In Equations 21, 22, and 23, there are six growth parameters, namely, $Y_{e/l}$, X, $Y_{e/X}$, μ, Z, and K_i, of which $Y_{e/l}$ and Y_{eX} can be combined into $Y = Y_{e/l}/Y_{e/X}$. There are also three environmental parameters $\emptyset I_o$, A, and V. To avoid photoinhibition at a particular radiance ($\emptyset I_o$), Y and A need to be decreased, or X, μ, Z, K_i, and V need to be increased. These effects can be achieved through the selection of appropriate strains and appropriate reactor design. It should be noted that the direct consequence of an increase in the values of X and V and a decrease in A is the creation of a larger dark fraction in the culture system, with the effect on biomass productivity discussed in Section IV.B. The photodynamic model proposed here has yet to be confirmed experimentally.

VII. CONCLUDING REMARKS

In general, growth kinetics models developed for bacterial cultures are applicable to photosynthetic cultures under ideal growth conditions, where each cell in the photosynthetic culture receives about the same amount of light energy over a short time interval. However, under most circumstances, light energy is not homogeneously distributed in a photosynthetic culture system, and very often part of the culture receives no energy at all at any given time. In extreme cases, cells next to the illuminated surface may suffer from photodynamic effects of high incident light intensity, while cells at the center of the culture may be completely starved of an energy supply. Kinetic modeling of photosynthetic cultures is admittedly very complicated. Nonetheless, with the growing understanding of culture physiology and bioenergetics, some realistic physiological models are being developed. The ultimate challenge in this area is modeling photosynthetic cultures subject to the transient kinetics resulting from diurnal intermittent illumination of varying intensities. There are still very few reported works on transient behavior of photosynthetic cultures in changing light intensity. There is also no well-tested physiological model of photoinhibition at high visible light intensities.

The construction of physiological models of photosynthetic cultures is important for several reasons. First, it enables a quantitative prediction of culture behavior under different environmental conditions. Second, such models once constructed provide insight into the ecology of phytoplankton, and finally, they allow for better control for the mass cultivation of photosynthetic cells for the production of products of potential commercial importance.

VIII. SYMBOLS AND UNITS

A	=	Illuminated culture surface area (m^2)
a	=	Absorptivity of liquid sample (m^2/kg)
C	=	Biomass carbon loss rate in dark (/sec)
D	=	Dilution rate (/sec)
E	=	Energy content of biomass (mJ^{-3})
F	=	Medium feed rate (m^3/sec)
I^K	=	Saturation light intensity ($J/sec/m^2$)
I_o	=	Incident light intensity ($J/sec/m^2$)
K_I	=	Saturation constant of light ($J/sec/m^2$)
K_i	=	Inhibition constant ($J/kg/sec$)
L	=	Light path (m)
M	=	Biomass of a cell (kg)
m	=	Maintenance energy coefficient ($J/kg/sec$)
m_r	=	Maintenance energy coefficient of resting cells ($J/kg/sec$)
P	=	Pigment content (kg/kg)
P_{max}	=	Maximum pigment content (kg/kg)
P_{min}	=	Minimum pigment content (kg/kg)

q	=	Specific rate of energy uptake (J/kg/sec)
q_e	=	Specific rate of accumulation of an energy intermediate (J/kg/sec)
r	=	Cell radius (m)
V	=	Total culture volume (m^3)
X	=	Biomass concentration (kg/m^3)
Y	=	Biomass yield from light utilized (kg/J)
$Y_{e/l}$	=	Production yield of an energy intermediate on light energy absorbed (J/J)
$Y_{e/X}$	=	Growth yield of biomass on the energy intermediate (g/J)
Y_G	=	Maximum growth yield from light utilized (kg/J)
Z	=	Nongrowth linked capacity in dissipating the energy intermediate (J/kg/sec)
α	=	Illuminated fraction of total culture volume (dimensionless)
β	=	Dark growth zone, a fraction of the total culture volume where cells continue to grow in dark (dimensionless)
δ	=	Dark fraction of the culture volume where cell growth is arrested
ø	=	Fraction of photosynthetically active radiance (dimensionless)
ρ	=	Density of cell mass (kg/m^3)
μ	=	Specific growth rate (/sec)
$μ_e$	=	Specific maintenance rate (/sec)
$μ_m$	=	Maximum specific growth rate (/sec)
$μ_o$	=	Overall (apparent) specific growth rate (/sec)

REFERENCES

1. **Burlew, J. S.,** Current status of the large-scale culture of algae, in *Algal Culture, from Laboratory to Pilot Plant,* Burlew, J. S., Ed., Carnegie Institution, Washington, D.C., 1953, 3.
2. **Benemann, J. R., Weissman, J. C., Koopman, B. L., and Oswald, W. J.,** Energy production by microbial photosynthesis, *Nature (London), 268,* 19, 1977.
3. **Pirt, S. J., Lee, Y. K., Richmond, A., and Watts Pirt, M.,** The photosynthetic efficiency of *Chlorella* biomass growth with reference to solar energy utilization, *J. Chem. Tech. Biotechnol., 30,* 25, 1980.
4. **Vincent, W. A.,** Algae and lithotrophic bacteria as food sources in, *Microbes and Biological Productivity,* Hughes, D. E. and Hughes, A. H., Eds., Cambridge University Press, London, 1971, 47.
5. **Richmond, A. and Preiss, K.,** The biotechnology of alga culture, *Interdisc. Sci. Rev.* 5, 60, 1980.
6. **Oswald, W. J.,** Productivity of algae in sewage disposal, *Solar Energy,* 15, 107, 1973.
7. **Rhee, G. Y.,** Continuous culture in phytoplankton ecology, in *Advances in Aquatic Microbiology,* Vol. 2, Droop, M. R. and Jannasch, H. W., Eds., Academic Press, New York, 1980, 151.
8. **Rhee, G. Y., Gothan, I. J., and Chisholm, S. W.,** Use of cyclostat cultures to study phytoplankton ecology, in *Continuous Cultures of Cells,* Vol. 2, Calcott, P. H., Eds., CRC Press, Boca Raton, Fla., 1981, 159.
9. **Cunningham, A. and Maas, P.,** The growth dynamics of unicellular algae, in *Microbial Population Dynamics,* Bazin, M. J., Ed., CRS Press, Boca Raton, Fla., 1982, 167.
10. **Myers, J.,** Growth characteristics of algae in relation to the problems of mass culture, in *Algal Culture, from Laboratory to Pilot Plant,* Burlew, J. S., Ed., Carnegie Institution, Washington, D.C., 1953, 37.
11. **Van Oorschot, J. L. P.,** Conversion of light energy in algal cultures, *Med. Van. Lund. Want., Nederland,* 55, 525, 1955.
12. **Shelef, G., Oswald, W. J., and Goleuke, C. G.,** Kinetics of Algal Systems in Waste Treatment: Light Intensity and Nitrogen Concentration as Growth-Limiting Factors, SERL Rep. No. 68.4., Sanitary Engineering Research Laboratory, University of California, Berkeley, 1968.
13. **Goldman, J. C.,** Outdoor algal mass culture. V. Photosynthetic yield limitations, *Water Res.,* 13, 119, 1979.
14. **Göbel, F.,** Quantum efficiencies of growth, in *The Photosynthetic Bacteria,* Clayton, K. R. and Sistrom, W. R., Eds., Plenum Press, New York, 1978, 907.
15. **Van Lieve, L. and Mur, L. R.,** Growth kinetics of *Oscillatoria agardhii* Gomont in continuous culture, limited in its growth by the light energy supply, *J. Gen. Microbiol.,* 115, 153, 1979.

16. **Ogawa, T. and Aiba, S.,** Bioenergetic analysis of mixotrophic growth in *Chlorella vulgaris* and *Scenedesmus acutus, Biotechnol. Bioeng.,* 23, 1121, 1981.

17. **Pirt, S. J.,** *Principles of Microbe and Cell Cultivation,* Blackwell Scientific, Oxford, 1975.

18. **Richardson, K., Beardall, J., and Raven, J. A.,** Adaptation of unicellular algae to irradiance: an analysis of strategies, *New Phytol.,* 93, 157, 1983.

19. **Pirt, S. J.,** The maintenance energy of bacteria in growing cultures, *Proc. R. Soc. London Ser. B:,* 163, 224, 1965.

20. **Pirt, S. J., Lee, Y. K., Walach, M. R., Watts Pirt, M., Balyuzi, H. H. M., and Bazin, M. J.,** A tubular bioreactor for photosynthetic production of biomass from carbon dioxide: design and performance, *J. Chem. Tech. Biotechnol.,* 33B, 35, 1983.

21. **Mituya, A., Nyunoya, T., and Tamiya, H.,** Pre-pilot-plant experiments on algal mass culture, in *Algal Culture, from Laboratory to Pilot Plant,* Burlew, J. S., Ed., Carnegie Institution, Washington, D. C., 1953, 273.

22. **Lee, Y. K., and Pirt, S. J.,** Energetics of photosynthetic algal growth: influence of intermittent illumination in short (40s) cycles, *J. Gen. Microbiol.,* 124, 43, 1981.

23. **Pipes, W. O. and Koutoyannis, S. P.,** Light-limited growth of *Chlorella* in continuous culture, *Appl. Microbiol.,* 10, 1, 1962.

24. **Eppley, R. W. and Dyer, W. L.,** Predicting production in light-limited continuous culture of algae, *Appl. Microbiol.,* 13, 833, 1965.

25. **Dabes, J. N., Wilk, C. R., and Sauer, K. H.,** The Behavior of *Chlorella pyrenoidosa* in Steady State Continuous Culture, UCRL-19958, Lawrence Radiation Laboratory, University of California, Berkeley, 1970.

26. **Myers, J.,** Prediction and measurement of photosynthetic production, in *Prediction and Measurement of Photosynthetic Production,* Centre for Agricultural Publication and Documentation, Wageningen, The Netherlands, 1970.

27. **Gons, H. J. and Mur, L. R.,** An energy balance for algal populations in light-limiting conditions, *Verh. Int. Ver. Limnol.,* 19, 2729, 1975.

28. **Richmond, A.,** Phototrophic microalgae, in *Biotechnology,* Vol. 3, Dellweg, H., Ed., Verlag Chemie, Weinheim, West Germany, 1983, 107.

29. **Watts Pirt, M. and Pirt, S. J.,** Photosynthetic production of biomass and starch by *Chlorella* in chemostat culture, *J. Appl. Chem. Biotechnol.,* 27, 643, 1977.

30. **Kok, B.,** Experiments on photosynthesis by *Chlorella* in flashing light, in *Algal Culture, from Laboratory to Pilot Plant,* Burlew, J. S., Ed., Carnegie Institution, Washington, D.C., 1953, 63.

31. **Lee, Y. K. and Pirt, S. J.,** Maximum photosynthetic efficiency of biomass growth, a criticism of some measurements, *Biotechnol. Bioeng.,* 24, 507, 1982.

32. **Lee, Y. K., Tan, H. M., and Hew, C. S.,** The effect of growth temperature on the bioenergetics of photosynthetic algal culture, *Biotechnol. Bioeng.,* 27, in press, 1985.

33. **Gons, H. J. and Mur, L. R.,** Energy requirements for growth and maintenance of *Scenedesmus protuberans* Fritsch in light-limited continuous cultures, *Arch. Microbiol.,* 125, 9, 1980.

34. **Tamiya, H., Shibata, K., Sasa, T., Iwamura, T., and Morimura, Y.,** Effect of diurnally intermittent illuminaton on the growth and some cellular characteristics of *Chlorella,* in *Algal Culture, from Laboratory to Pilot Plant,* Burlew, J. S., Ed., Carnegie Institution, Washington, D.C., 1953, 76.

35. **Gons, H. J. and Mur, L. R.,** Growth of *Scenedesmus protubarans* Fritsch in light-limited continuous cultures with a light-dark cycle, *Arch. Hydrobiol.,* 85, 41, 1979.

36. **Myers, J. and Graham, J.,** On the mass culture of algae. II. Yield as a function of cell concentration under continuous sunlight irradiance, *Plant Physiol.,* 34, 345, 1959.

37. **Myers, J. and Graham, J.,** On the mass culture of algae. III. Light diffuses: light vs low temperature Chlorellas, *Plant Physiol.,* 36, 342, 1961.

38. **Laws, E. A. and Bannister, T. T.,** Nutrient and light-limited growth of *Thalassiosira fluviatilis* in continuous culture, with implications for phytoplankton growth in the ocean, *Limnol. Oceanogr.,* 25, 547, 1980.

39. **Walach, M. R., Balyuzi, H. M., Bazin, M. J., Lee, Y. K., and Pirt, S. J.,** Computer control of an algal bioreactor with simulated diurnal illumination, *J. Chem. Tech. Biotechnol.,* 33B, 59, 1983.

40. **Walach, M. R., Balyuzi, H. H. M., Lee, Y. K., and Pirt, S. J.,** Computer control of photobioreactor to maintain constant biomass during diurnal variation in light intensity, in *Microbiological Methods for Environmental Biotechnology,* Grainger, J. M. and Lynch, J. M., Eds., Academic Press, London, 1984, 313.

41. **Goldman, C. R., Mason, D. T., and Wood, B.,** Light injury and inhibition in antarctic fresh water phytoplankton, *Limnol. Oceanogr.,* 13, 313, 1963.

42. **Krinsky. N. I.,** Cellular damage initiated by visible light, in *The Survival of Vegetative Microbes,* Gray, T. R. G. and Postgate, J. R., Eds., Cambridge University Press, London, 1976, 209.

43. **Belay, A. and Fogg, G. E.,** Photoinhibition of photosynthesis in *Asterionella formosa* (Bacillariophyceae), *J. Phycol.,* 14, 341, 1978.
44. **Harris, G. P.,** Photosynthesis, productivity and growth: the physiological ecology of phytoplankton, *Ergeb. Limnol.,* 10, 1, 1978.
45. **Knoechel, R. and Kaliff, J.,** An in situ study of the productivity and population dynamics of five fresh water planktonic diatom species, *Limnol. Oceanogr.,* 23, 195, 1978.
46. **Spikes, J. D.,** Photodynamic action, in *Photophysiology.* Vol. 3, Evese, A. C., Ed., Academic Press, New York, 1968, 33.
47. **Spikes, J. D. and Livington, R.,** The molecular biology of photodynamic action sensitized oxidations in biological system, *Adv. Radiat. Biol.,* 3, 29, 1969.
48. **Kaplan, A.,** Photoinhibition in *Spirulina platensis:* response of photosynthesis and HCO_3 uptake capability to CO_2 depleted conditions, *J. Exp. Bot.,* 32, 669, 1981.
49. **Gerber, D. W. and Burris, J. E.,** Photoinhibition and p700 in the marine diatom *Amphora* sp., *Plant Physiol.,* 68, 699, 1981.
50. **Vermeglio, A. and Carvier, J. M.,** Photoinhibition by flash and continuous light of oxygen uptake by intact photosynthetic bacteria, *Biochem Biophys. Acta,* 764, 233, 1984.

Chapter 6

ROLE OF pH IN BIOLOGICAL WASTEWATER TREATMENT PROCESSES

Prasad S. Kodukula, T. B. S. Prakasam, and Arthur C. Anthonisen

TABLE OF CONTENTS

I. INTRODUCTION

The term pH is used to describe the concentration of hydrogen ions in aqueous and other liquid media. pH is an important factor in wastewater treatment processes because of its influence on the process performance and efficiency. It plays an important role in both chemical and biological wastewater treatment systems. In the case of chemical unit operations, such as neutralization and precipitation, the treatment objective is achieved solely by adjusting the pH of the wastewater under consideration. The reaction rates in these systems are usually much faster than in biological wastewater treatment systems. In biological wastewater treatment systems, however, although the adjustment of pH is not the primary means of achieving the treatment objective, the process efficiency is nevertheless influenced by the pH, mainly because of the effect of the hydrogen ion concentration on the metabolic activity of the responsible microoogranisms.

In the case of some chemical treatment processes, the pH of the influent is adjusted to the desired level by addition of appropriate chemicals. Examples of such processes are (1) precipitation of phosphates[1] and heavy metals[2] (2) neutralization of several industrial wastes by addition of either an acid or an alkali to the process stream, depending on the inital pH of the waste, or by mixing wastes having pH values in the acidic and alkaline range[3] (3) lime stabilization of sludges to destroy bacteria and viruses[4], and (4) hydrolysis of organic wastes by raising or lowering their pH through alkali or acid treatment, respectively, to increase their biodegradability.[5] In the processes of adsorption[6] and coagulation/flocculation,[7] in which the adjustment of pH is not necessarily the means of treatment, variations in pH may alter the ionic charges of different chemical groups in the system, thereby affecting the process efficiency.

In biological wastewater treatment processes, although the adjustment of pH to an optimum level is not a means to achieve the stabilization of carbonaceous matter directly, the provision of such optimal pH conditions is conducive to the growth of microorganisms that are responsible for its stabilization.

In a process such as chlorination of effluents, the treatment objective is disinfection. Although the pH of wastewaters is not usually manipulated for achieving disinfection, the efficacy of chlorine as a disinfecting agent is nevertheless influenced by the pH of the wastewater. The relative concentrations of the different species of chlorine formed at various pH values have different germicidal efficiencies.[8]

Biological treatment systems, the focus of this chapter, may be influenced by hydrogen ion concentration in different ways. Different groups of microorganisms may become dominant at different pH levels. For example, acidic conditions generally favor the growth of fungi, and so they form the predominant group of microorganisms under such conditions.[9] Furthermore, the activity of microorganisms may be inhibited under unfavorable pH conditions, although other environmental factors are satisfactory for their growth.

While extremely low or high hydrogen ion concentrations may be directly inhibitory to the enzymatic activity of microorganisms contained in wastewater treatment systems, inhibition of their activity may also be caused by certain chemical species, whose ionic state is altered as a result of a change in the pH of the system. For example, ammonia exists as un-ionized or free ammonia (FA) rather than ammonium ion in increasing concentrations as the pH is progressively increased in the alkaline range. A high concentration of FA, not necessarily ammonium ions, is known to inhibit several biological processes including nitrification[10] and anaerobic digestion.[11] Also, pH affects the equilibria of sulfide and sulfite species. Hydrogen sulfide, an odorous gas, is in equilibrium with sulfides and is the predominant species at low pH levels. It is also known to be lethal to humans at high concentrations.[12] The bisulfite ion (HSO_3^-) produced by the dissociation of H_2SO_3 at pH values in the alkaline range is toxic to fish.[13] Similarly, pH determines the speciation of heavy metals in sludge-amended soils which in turn, influences the degree of environmental hazard associated with crops grown on such soils.[14]

In addition to playing a key role in influencing the performance of biological waste treatment processes, extremely low and high hydrogen ion concentrations adversely affect the aquatic life in receiving waters. In order to assure a satisfactory environment for such life forms, most of the states have regulatory pH standards for municipal as well as industrial effluents that are intended for discharge into receiving waters.

II. CHEMISTRY

The term pH is defined as follows:

$$pH = -\log[H^+] \tag{1}$$

where $[H^+]$ is the concentration of hydrogen ions in terms of mol/ℓ. In the strictest sense, activity of hydrogen ions should be used in the definition, however, for practical purposes, molar concentrations approximate activities. The pH scale ranges from 0 to 14 with pH 7 representing neutrality. Those pH values which are less than 7 are acidic, and those greater than 7 are alkaline.

A. Acid-Base Equilibria

According to commonly used Bronsted-Lowry theory, an acid is defined as a proton donor and a base as a proton acceptor. As a general rule, any species (for example, HA) that can donate a proton, on dissociation, gives a species (A^-) that can accept a proton to form HA, as shown in the following equilibrium reaction:

$$HA \rightleftharpoons H^+ + A^- \tag{2}$$

In the Bronsted-Lowry system, the above reaction is considered to involve a proton transfer from HA to a molecule of solvent:

$$HA + H_2O \rightleftharpoons H_3O^+ + A^- \tag{3}$$

Every Bronsted-Lowry acid is paired with a Bronsted-Lowry base and the two are said to comprise a conjugate acid-base pair. For example, HCl is the conjugate acid of Cl^-; Cl^- is the conjugate base of HCl.

An acid such as HCl dissociates completely in solution into H^+ and Cl^- ions, and hence is considered a strong acid. On the other hand, a weak acid such as acetic acid dissociates only partly, being in equilibrium with its representative ions. For example, the undissociated acetic acid is in equilibrium with acetate ions in a solution as described by the following reaction:

$$CH_3COOH \rightleftharpoons CH_3COO^- + H^+ \tag{4}$$

Similarly, strong bases and weak bases can be defined based on the degree of their ionization in a solution.

B. Buffers

For wastewater biological treatment processes, it is important that the pH not deviate significantly from an optimum value that is suited. A nearly constant pH may be maintained due to what is called the buffering action of acid-base equilibria and can be achieved by the addition of buffering compounds, if needed, or as a result of natural chemical reactions such as those involving carbonates and bicarbonates (discussed in a later section). A buffer contains both an acid and a base, and can respond to the addition of either an acid or alkali, with a minimal change in pH. As an example, a solution containing acetic acid molecules and acetate ions in equilibrium (as represented in Equation 4) is considered a buffer as shown below.

The following equilibrium can be obtained for the reaction in Equation 4:

$$K_a = \frac{[CH_3COO^-][H^+]}{[CH_3COOH]} \tag{5}$$

where [] indicate molar concentrations and K_a is the thermodynamic equilibrium constant for dissociation of acetic acid.

Solving for $[H^+]$,

$$[H^+] = \frac{K_a[CH_3COOH]}{[CH_3COO^-]} \tag{6}$$

Taking the negative logarithm on both sides,

$$-\log[H^+] = -\log K_a - \log\frac{[CH_3COOH]}{[CH_3COO^-]} \tag{7}$$

or

$$pH = pK_a + \log\frac{[CH_3COO^-]}{[CH_3COOH]} \tag{8}$$

The negative log of K_a is now designated pK_a (a designation analogous to that of log $[H^+]$). The pK_a of acetic acid is approximately 4.7 at 25°C. Equation 8 is known as the Henderson-Hasselbach equation, and is applicable in biological systems, where the pH has to be maintained within a narrow range.

In a solution where the ratio of $[CH_3COO^-]/CH_3COOH]$ (i.e., dissociated ion to undissociated acid) is equal to unity, pH is equal to pK_a, since log of 1.0 is equal to 0. If a small amount of strong acid is added to such a solution, some of the acetate ions are converted to acetic acid; if a strong base is added, a portion of the acetic acid is converted to acetate ions. In either case, the change in the ratio of $[CH_3COO^-]/[CH_3COOH]$ is very small, and the subsequent change in pH is even less. This pH change would be very small compared to that produced in the absence of the buffer. Thus, a mixture of acetic acid and acetate ions is considered a buffer for pH of 4.7.

It can also be seen from Equation 8 that a pH higher than pK_a, the dominating species would be the dissociated ion, whereas at a pH less than the pK_a, the undissociated acid would form the major species.

C. Chemical Equilibria Biological Systems

The major chemical equilibria, which are of significance in biological wastewater treatment systems include: carbonate, ammonia, sulfide, and chlorine. The chemical species that make up each one of these systems are interrelated within the given system, and the relative concentrations of these species are primarily a function of hydrogen ion concentration. Table 1 presents chemical equilibria and equilibrium constants for all the reactions involving different chemical species of each system.

Table 1
THERMODYNAMIC
EQUILIBRIUM
CONSTANTS FOR
DIFFERENT
EQUILIBRIUM
REACTIONS

Equilibrium reaction	pK$_a$
$CO_{2(aq)} + H_2O \rightleftharpoons H_2CO_3$	2.8
$H_2CO_3 \rightleftharpoons HCO_3^- + H^+$	6.3
$HCO_3^- \rightleftharpoons CO_3^{2-} + H^+$	10.3
$NH_4^+ \rightleftharpoons NH_3 + H^+$	9.3
$H_2S \rightleftharpoons HS^- + H^+$	7.1
$HS^- \rightleftharpoons S^{2-} + H^+$	14.0
$HOCl \rightleftharpoons OCl^- + H^+$	7.5

Based on the information presented in Table 1, the relative concentrations of different species in a given system can be calculated as a function of pH. Since acid-base reactions such as those presented in Table 1 are usually very rapid, equilibrium calculations can be used to determine concentrations of different species of the acid and its conjugate base and of the hydrogen and hydroxyl ions at a given time. Such calculations, however, are tedious, particularly when dealing with multiprotic acids, and so graphical procedures have been developed to determine concentrations of different species in any given acid-base system. Such procedures include pC-pH diagrams and distribution diagrams. The former involve setting up equilibrium and mass balance equations which are translated to plots of $-\log$ concentration (pC) of each species involved vs. the pH. A pC-pH diagram so produced can be used to first determine the pH where all the charges involved are balanced. (This is referred to as proton condition.) All of the species concentrations can then be read off the diagram at this pH value. As an example, a pC-pH diagram of a 10^{-3} *M* solution of H$_2$A (pK$_{a1}$ = 4.0; pK$_{a2}$ = 8.0) is presented in Figure 1.

In order to develop a concentration diagram, a total analytical concentration C$_t$ of H$_2$A must be specified; thus, in Figure 1, C$_t$ is fixed to be 10^{-3}. The H$^+$ and OH$^-$ concentrations are represented by straight lines with slopes equal to -1 and 1, respectively. A change in C$_t$ would shift the curves in Figure 1 vertically with virtually no change in their shapes and relative positions.

The H$_2$A and HA$^-$ curves in Figure 1 intersect at a concentration of ($^1/_2$)C$_t$, which is 0.3 log units (i.e., $\log_{10}$ 2) below point A where pH = pK$_{a1}$. Similarly, the HA$^-$ and A^{2-} curves intersect at 0.3 log units below point B where pH = pK$_{a2}$. These points are referred to as "system points", where equal concentrations of the two species involved are present, and so the most buffering action is achieved. In Figure 1, the system points are

$$\text{Point A:} \quad H_2A \rightleftharpoons HA^- + H^+; \quad pH = 4.0$$

$$\text{Point B:} \quad HA^- \rightleftharpoons A^{2-} + H^+; \quad pH = 8.0$$

In developing pC-pH diagrams from mass balance equations, several assumptions are made, which break down in pH regions close to pK$_2$ values. Furthermore, the equilibrium calculations become increasingly complex as the number of protons in the acid increase. However, graphical solutions of equilibrium problems may be simplified by use of the so-called distribution diagrams. These diagrams are graphical presentations of relationships

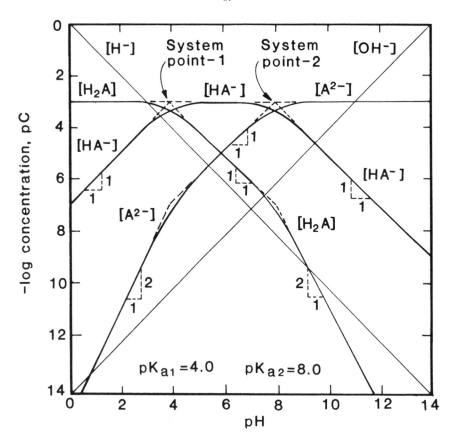

FIGURE 1.   pC-pH diagram for H_2A.

between the pH of a given system and the amount of species in the system as a fraction of the total analytical concentration. These ionization fractions, which are called α values, can be determined from the equilibrium and mass balance equations. The α for a diprotic acid such as H_2A are represented by

$$\alpha_0 = \frac{[H_2A]}{C_t} \tag{9a}$$

$$\alpha_1 = \frac{[HA^-]}{C_t} \tag{9b}$$

$$\alpha_2 = \frac{[A^{2-}]}{C_t} \tag{9c}$$

where C_t is the total analytical concentration of the species under consideration. The subscript α refers to the number of protons that the species has released. The α value is independent of the total analytical concentration, and therefore, multiplication of the analytical concentration by the appropriate α value at the particular pH will directly produce the concentration of the species at that pH. Distribution diagrams as well as pC-pH diagrams for different acid-base systems are presented in the following sections under different headings. For a detailed discussion on pC-pH and distribution diagrams, the reader is referred to References 15 and 16.

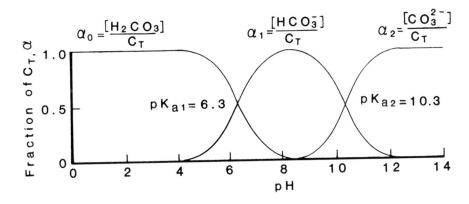

FIGURE 2.   Distribution diagram for dissolved carbonates at 25°C.

1. Carbonate

The carbonate system includes the species of gaseous carbon dioxide (CO_{2g}), dissolved carbon dioxide (CO_{2aq}), carbonic acid (H_2CO_3), bicarbonate (HCO_3^-), and carbonate-containing solids. The carbonate system is important in biological systems for at least two reasons: (1) the carbonate and bicarbonate species act as buffers, and thus assist in moderating any sudden shifts in pH and (2) CO_2 may be introduced into the system by atmospheric dissolution or biological processes such as respiration and biodegradation, or removed from the system by photosynthesis or synthesis of autotrophs. (Carbon dioxide is the sole carbon source for the autotrophic organisms, and its availability is affected by pH).

The chemical equilibria and equilibrium constants for reactions involving carbonate species are presented in Table 1. Carbonate containing solids in precipated form are ignored here because of their relative lack of significance in wastewater systems. However, carbon dioxide may evolve in significant quantity from these solids, if the pH turns highly acidic. Since it is difficult to distinguish analytically between dissolved CO_2 and carbonic acid, they are, as a sum, designated as hypothetical $H_2CO_3^*$. Furthermore, it is generally assumed that concentrations of H_2CO_3 and CO_{2aq} in aqueous systems are equal.

A distribution diagram and a concentration diagram for dissolved carbonates in an aqueous system that is not in contact with CO_{2g} are presented in Figures 2 and 3, respectively.

If an aqueous system is in contact with a gas phase containing carbon dioxide, Figure 3 is no longer applicable. Figure 4 presents a concentration diagram for an open system in contact with gaseous carbon dioxide with a partial pressure of 3×10^{-4} atm. In an open system, as pH is increased, $H_2CO_3^*$ is dissociated, but in order to maintain equilibrium, more gaseous CO_2 enters the solution, thus keeping the concentration of $H_2CO_3^*$ constant (see Figure 4). However, in a closed system, as pH increases, $H_2CO_3^*$ dissociates, and its concentration decreases, as no gaseous CO_2 enters the system. Furthermore C_t increases with pH in open systems due to CO_2 dissolution, but remains constant in closed system.

2. Ammonia

Ammonia equilibrium is of considerable interest in the area of biological treatment because of the toxicity associated with FA. Biological processes, such as nitrification and anaerobic digestion are particularly sensitive to FA toxicity. Distribution of ammonia between the dissociated and undissociated forms is primarily a function of pH as shown in Figures 5 and 6. At pH levels above 9.5, FA is the dominating species.

3. Sulfide

Sulfides are of considerable concern in wastewater treatment for two reasons: odor prob-

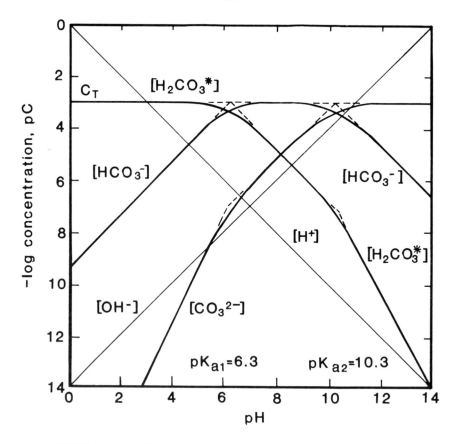

FIGURE 3.   pC-pH diagram for dissolved carbonated in a closed system at 25°C.

lems and sewer corrosion. These problems occur as a result of reduction of sulfate to sulfide under anaerobic conditions. In the absence of dissolved oxygen and nitrates, sulfates serve as hydrogen acceptors for biochemical reactions involving the anaerobic decomposition of organic matter and are reduced to sulfide, which is in equilibrium with hydrogen ions as shown in Table 1. The relationships between H_2S, HS^-, and S^{2-} at various pH levels are shown in Figures 7 and 8 in the form of pC-pH and distribution diagrams, respectively. As can be seen from these figures, the concentration of H_2S is negligible at pH values of about 8 and above. However, at pH values of lower than 8, the equilibrium is shifted progressively toward the formation of H_2S, which would result in an increase of the partial pressure of H_2S in the vapor phase. It is the vapor phase H_2S that causes serious odor problems.

Hydrogen sulfide is also responsible for causing "crown" corrosion of concrete sewers. In domestic sewers, sulfates may be reduced by bacteria such as *Desulfovibrio dusulfuricans* to sulfides under anaerobic conditions. At typical pH levels of domestic sewage, most of the sulfides exist as hydrogen sulfide, a fraction of which may escape into the atmosphere above the sewage. The H_2S gas present in the vapor phase may be oxidized to sulfuric acid by sulfur oxidizing bacteria such as *Thiobacillus concretivorus* which can function at a sulfuric acid concentration of at least 7%. The sulfuric acid thus formed in sewers may result in corrosion of the concrete walls of the sewer lines. A schematic sketch of the reactions involved in crown corrosion of concrete sewers is presented in Figure 9.

4. Chlorine

When chlorine gas is dissolved in water, it hydrolyzes rapidly according to the equation:

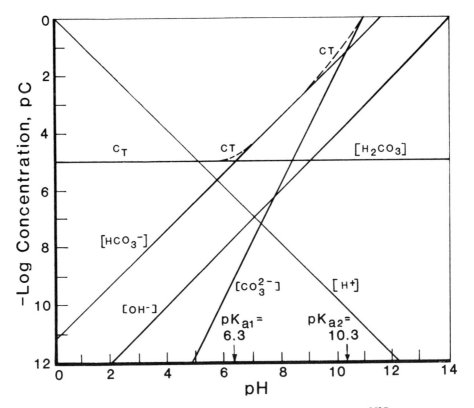

FIGURE 4. pC-pH diagram for carbonates in an open system at 25°C.

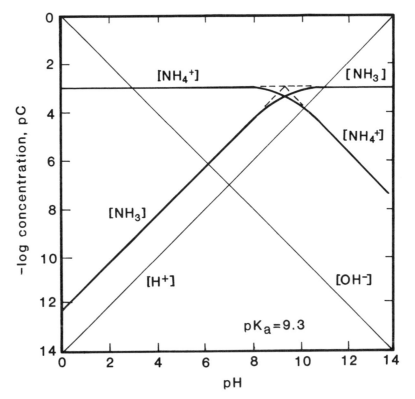

FIGURE 5. pC-pH diagram for ammonia at 25°C.

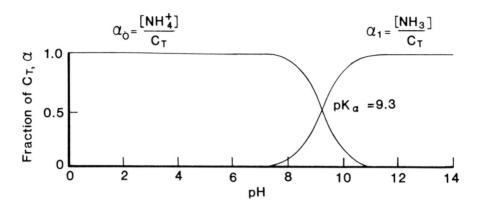

FIGURE 6.   Distribution diagram for ammonia at 25°C.

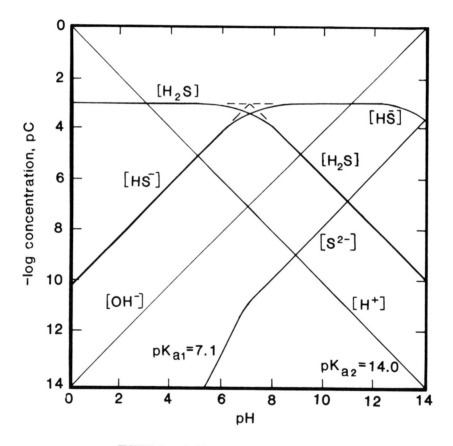

FIGURE 7.   pC-pH diagram for sulfides at 25°C.

$$Cl_2 + H_2O \rightarrow H^+ + Cl^- + HOCl \tag{10}$$

HOCl is hypochlorous acid, and being a weak acid, undergoes partial dissociation as follows:

$$HOCl \rightleftharpoons H^+ + OCl^- \tag{11}$$

to produce hypochlorite and hydrogen ions. Chlorine, hypochlorite ion, and hypochlorous

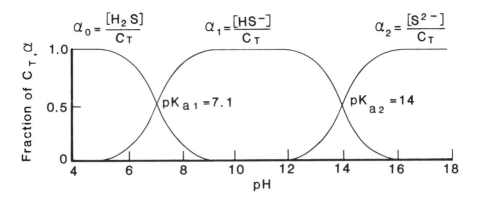

FIGURE 8. Distribution diagram for sulfides at 25°C.

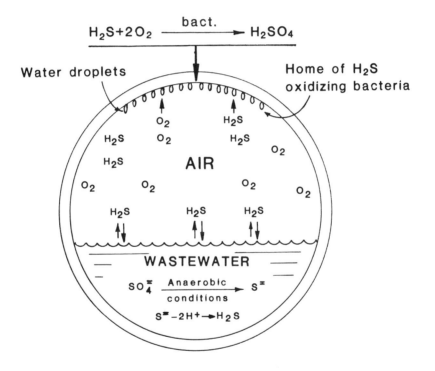

FIGURE 9. Schematic representation of crown corrosion in concrete sewers.[12]

acid are referred to as free chlorine residual. The ionization constant for HOCl is given in Table 1, and the pC-pH and distribution diagrams for HOCl are presented in Figures 10 and 11, respectively. It can be seen from these figures that at pH levels greater than 7.5, the dominating species is OCl^- ion.

In dilute aqueous solutions, HOCl reacts with ammonia-forming chloramines, which are referred to as combined chlorine residual.

$$HOCl + NH_3 \rightarrow NH_2Cl \text{ (monochloramine} + H_2O \qquad (12)$$

$$NH_2Cl + HOCl \rightarrow NHCl_2 \text{ (dichloramine)} \quad + H_2O \qquad (13)$$

$$NHCl_2 + HOCl \rightarrow NCl_3 \text{ (trichloramine)} \qquad + H_2O \qquad (14)$$

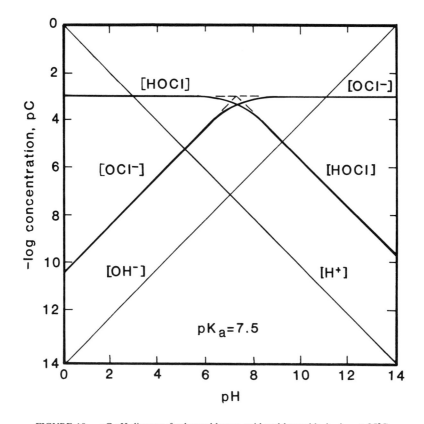

FIGURE 10.   pC-pH diagram for hypochlorous acid and hypochlorite ion at 25°C.

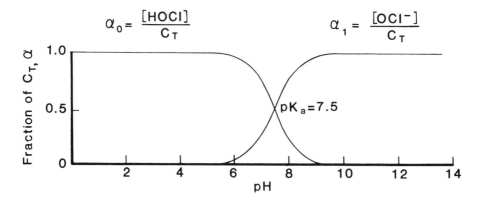

FIGURE 11.   Distribution diagram for hypochlorous acid and hypochlorite ion at 25°C.

The above reactions are in general by steps, and so they will compete with each other. Hence, the rate of the first reaction is extremely important, and has been shown to be pH sensitive.[8] The fastest conversion of HOCl to NH_2Cl has been reported to occur at a pH of 8.3. The pH dependence of this reaction can be described on the basis of the HOCl-OCl⁻ and NH_4^+-NH_3 equilibria. The equilibrium condition among the chloramines is a function of pH. At pH levels of about 9 or above monochloramines exist almost exclusively; at pH about 6.5, mono- and dichloramines coexist in approximately equal amounts; below pH 6.5, dichloramines predominate; and trichloramines exist below pH about 4.5. The progressive addition of chlorine to water or wastewater containing ammonia results initially in an increase in the concentration of the combined chlorine residual and then in a decrease of the same to a minimum, beyond which the added chlorine is in the form of free residual chlorine. Chlorination up to the point where a minimum (or zero level) of combined chlorine exists is termed break point chlorination. In the break point chlorination of water and wastewater, nitrogenous compounds other than NCl_3 are also known to be formed. These include N_2, N_2O, and NO_3^-. The relative distribution of these compounds is governed by the pH of the medium, NCl_3 being the preponderant species at very low pH values.[8]

III. ALTERATION OF pH BY MICROBIAL ACTIVITY

In most biological systems, acidic or alkaline products are produced due to metabolic activity of microorganisms. Production of such substances may result in the alteration of the pH of the surroundings, depending upon their concentration and the buffering capacity of the system.

Oxidation of organic substances under aerobic conditions generally produces carbon dioxide, which may lower the pH. In anaerobic systems, however, acidic intermediate products are formed by what are known as "acid formers", and under favorable conditions, are converted to methane and carbon dioxide. The acid intermediates may sometimes accumulate in the reactor and reduce the pH, depending upon the alkalinity of the system, thereby inhibiting the overall anaerobic process. Even under aerobic conditions, some bacteria such as ammonia oxidizers produce acidic metabolic intermediates, which may decrease the pH of the system.

Whereas several types of bacteria, as in the cases discussed above, form acidic intermediates or end products, which may or may not be inhibitory to the growth of such bacteria, a few specialized groups of bacteria produce a large amount of acid resulting in a pH level as low as 2, which is favorable for the growth of these bacteria. For example, several species of *Thiobacillus* oxidize sulfur, sulfide, and thiosulfate to form sulfuric acid. Thus, by sulfur oxidation, these organisms reduce the pH of their surroundings to less than 2. Sulfur oxidizing bacteria are unusually resistant to these acidic conditions, and are commonly found in acidic environments. These organisms have a significant role in at least two environmental situations. First, as described earlier, in concrete sewers, the H_2S gas produced from sewage under anaerobic conditions may be converted to H_2SO_4 by microorganisms living in the slimes attached to the inner wall of the sewer, causing the corrosion of the concrete. Second, in acid mine drainage waters, low pH conditions are produced due to the formation of H_2SO_4 from oxidation of sulfides in iron ore. On the other hand, the ability of sulfur oxidizing bacteria to produce acid is taken advantage of (1) to leach metals such as copper from low grade ores; (2) to release oil from shale by dissolving the rocky material, thus enabling the extraction of oil easier; and (3) to reduce the pH of alkaline soils to levels suitable for agricultural crops.

Alkaline products are also released by some microorganisms, causing an increase in pH of their surroundings. Deamination of nitrogenous organic compounds results in the formation of ammonia which can cause an increase in pH, when adequate buffering capacity is not present in the system. During the denitrification process, where the nitrates are converted to gaseous nitrogen, hydroxyl ions are released. In oxidation ponds and eutrophic water bodies with large amounts of algae, carbon dioxide is removed by algae for photosynthesis and consequently the pH of the surroundings is increased. pH levels of up to 11 have been reported in some oxidation ponds during active photosynthetic periods.[17]

Since the growth of microorganisms responsible for waste stabilization is possible only within a small range of pH values, it is essential to operate the waste treatment systems around an optimum pH. Although hydrogen or hydroxyl ions are generated due to microbial activity in many of the biological processes, the concomitant pH changes in these systems are usually minimal due to the available buffering action and the large dilution medium available. In biological treatment of sewage, pH is generally not a serious concern, because the pH of sewage is generally around neutral, and sewage contains sufficient buffering capacity due to the presence of amino acids and bicarbonates. On the other hand, for several industrial wastes, it is essential that their pH is adjusted to near neutrality for optimum biological treatment. Combining acidic wastes with alkaline wastes for their mutual neutralization is also practiced prior to biological treatment of industrial wastes.

In nitrification and dentrification systems, pH fluctuations may occur, and are of serious concern when they become extreme because of the sensitivity of the organisms to such variations. Similarly, in anaerobic treatment, pH control is important, because the methanogenic bacteria involved in the process have a very narrow range of pH tolerance.

The use of buffers to maintain a relatively constant pH in full-scale systems is not common. However, several inorganic buffers are used in laboratory and pilot-scale studies. The most commonly used buffer is a mixture of phosphate salts which provides a good buffering capacity near neutrality. In commercial systems, the pH is maintained at a desired level by automatic addition of acid or alkali as necessary.

IV. INHIBITION OF ENZYMATIC ACTIVITY

The hydrogen ion concentration is considered to be one of the most important factors that influences enzymatic activity. Each enzyme has a pH optimum where it reacts at the maximum speed. The rate of reaction decreases at higher or lower pH values than the optimum, and may be represented by a bell shaped pH vs. activity curve. Some enzymes have broad pH ranges while others have narrow ones. The pH optimum and the pH range within which a given enzyme is quite active varies sometimes radically with varying conditions of temperature, concentration of substrate, or other factors. For example, the lower the pH, the lower is the heat resistance of some bacterial spores.[18] Extreme variations in pH may result in complete inactivation of the enzyme. This reaction may be reversible, if the enzyme is not under extreme pH values for too long a time. The inactivation is usually accompanied by the formation of a denatured protein with a solubility differing from that of the active enzyme. In many cases the adjustment of pH to neutral levels would cause a reversal of denaturation, and the activity of the enzyme reappears. Irreversible inactivation or complete destruction of the enzyme is possible when the enzyme is subjected to long periods of extreme pH conditions. Since enzymes are responsible for catalyzing the reactions associated with life processes, it is reasonable to assume that the net effect of pH (or, for that matter, any of the influential factors) on various enzymes in a given organism is manifested in the growth response. Just as there is a pH optimum for the activity of each enzyme, there is also an optimum pH for the overall growth. This does not necessarily mean that the pH optimum for all the enzymes and the overall growth would be the same. It is possible to find in cells, that grow most rapidly at neutral pH levels, specific enzymes with pH optima at 9 or above. Thus, the optimum conditions for activity of one enzyme are not necessarily optimum for other enzymes or for the functioning of an entire system.

In addition to altering the activity of enzymes in cells, the hydrogen ion concentration also plays a key role in terms of enzymatic content. Cells of *Escherichia coli* grown in a medium at pH levels ranging from 4.5 to 9.5 were reported to produce different quantities of enzymes as shown in Figure 12. Although the enzymatic constitution, qualitatively or quantitatively, changes with the environmental conditions such as pH, it should not be misconstrued to mean that there is no consistent enzyme pattern for a given organism.

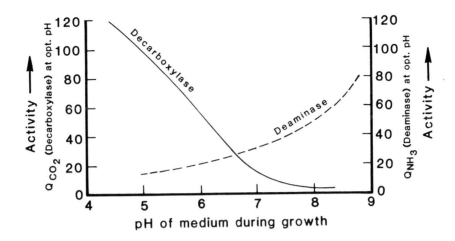

FIGURE 12.   Variations in the formation of glutamic acid decarboxylase and deaminase by *E. coli* with the pH of the medium during growth.[19]

V. MOLECULAR BASIS OF EFFECTS OF pH

It is postulated that the internal pH of a microbial cell is different from the outside pH, and it is likely that the pH may vary in localized areas within the cell.[20] Although microorganisms are found in habitats over a wide pH range, the internal pH of the cell is probably close to neutrality. The mechanisms through which the neutrality is maintained are not clearly understood. It is possible that the cells may keep hydrogen ions from entering inside, or actively expelling the ions as rapidly as they enter. Nevertheless, the difference in pH between the inside and outside the cell determines the mechanism through which the cellular activities are influenced by the hydrogen ion concentration.

A pH gradient is considered to exist across the cell membrane and is responsible for the driving force for the transport of nutrients and organic compounds into the cell.[21] It has also been shown that this pH gradient across the membrane is affected by the pH of the medium, thus influencing the chemical transport across the membrane.[21] The pH effects on the chemical transport may be direct or indirect. A direct effect would be by alteration of the pH gradient across the cell membrane. For example, pH influences the activity of specific membrane proteins, which bind different compounds for transport across the cell membrane. Indirect effects of pH may be caused by alteration of ionic state of nutrients, organics, or toxic compounds due to pH change. Non ionized compounds are more permeable through the cell membrane than ionized ones, and so depending on pH, their availability may change. While increased availability of nutrients may be an advantage to the organisms, enhanced transport of some may prove inhibitory and even toxic as a result of the alteration in pH of the environment. Examples of these include inhibition of nitritifiers (ammonia oxidizers), and nitratifiers (nitrite oxidizers) in the nitrification process[10] and methanogens in anaerobic systems[11] by non-ionized ammonia formed from ammonium ions at high pH values. Free nitrous acid is also detrimental to nitrifying organisms. The inhibition of nitrification by FA and nitrous acid is discussed further below. On the other hand, heavy metal ions, which are generally toxic at high concentrations, may become ineffective by precipitation at proper hydrogen ion concentration.[16] In land treatment systems, proper soil pH can make the heavy metals unavailable for plant uptake, thus reducing the health hazard associated with crops grown on sludge-amended soils. Furthermore, waste loadings may be increased on such soil systems due to reduced potential of heavy metal problems.[14]

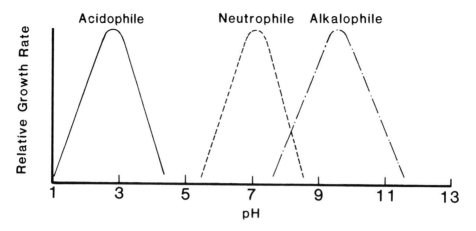

FIGURE 13.   Classification of microorganisms based on their affinity for the acidity of their environment.[22]

Table 2
MINIMUM, OPTIMUM, AND MAXIMUM pH
VALUES FOR SELECTED BACTERIA[23]

Organism	Minimum	pH optimum	Maximum
Escherichia coli	4.4	6.0—7.0	9.0
Proteus vulgaris	4.4	6.0—7.0	8.4
Aerobacter aerogenes	4.4	6.0—7.0	9.0
Pseudomonas aeruginosa	5.6	6.6—7.0	8.0
Erwinia carotovora	5.6	7.1	9.3
Clostridium sporogenes	5.0—5.8	6.0—7.6	8.5—9.0
Nitrosomonas spp.	7.0—7.6	8.0—8.8	9.4
Nitrobacter spp.	6.6	7.6—8.6	10.0
Thiobacillus thiooxidans	1.0	2.0—2.8	4.0—6.0
Lactobacillus acidophilus	4.0—4.6	5.8—6.6	6.8

VI. pH AND BIOLOGICAL WASTE TREATMENT

Biological wastewater treatment systems are mixed culture systems consisting of several groups of microorganisms with different physical and biochemical properties. Each group of microorganisms has a pH range within which growth is possible, and each group usually has a well-defined optimum pH. While most of the biological systems operate efficiently between pH values of 6 and 9 with optimum levels around neutrality, there are a few species that can grow at pH levels less than 2 or greater than 10. Based on their affinity toward the acidity of their environment, microorganisms can be classified into three different categories, namely acidophiles, neutrophiles, and alkalophiles. The ranges and the optimum pH values for these three groups of organisms are presented in Figure 13.

The maximum, minimum, and optimum pH values for several bacteria, generally found in aerobic systems are listed in Table 2. Although the range of pH units within which biological growth is possible seems to be narrow, it should be remembered that a difference of one pH unit represents a tenfold change in the hydrogen ion concentration. Table 3 presents minimum, maximum, and optimum pH values for different groups of microorganisms. Most bacteria grow best under neutral or slightly alkaline conditons; however, there are a few extremely acidophilic bacteria, such as *Thiobacillus thiooxidans* and some iron

Table 3
MINIMUM, MAXIMUM, AND
OPTIMUM pH VALUES FOR
DIFFERENT GROUPS OF
MICROOGANISMS

Group	Range	Optimum
Bacteria	5.0—9.0	7.0
Fungi	2.0—7.0	5.0
Blue green bacteria	6.0—9.0	>7.0
Protozoa	5.0—8.0	7.0

bacteria, which can grow at low pH values. Most yeasts and fungi prefer an acid environment. Many algae are tolerant of mild acidity with a few being extremely acidophilic. Most blue green bacteria have pH optima on the alkaline side of neutrality.

The effect of pH on biological processes may be studied under two different headings: effect of pH on previously acclimated systems and effect of pH shock loads. In the first case the microorganisms are adapted to a given pH condition, whereas in the second case, the pH of the system is subjected to an abrupt change by a shock load. There is very little information available in either case.

pH values for most domestic wastewaters are known to lie generally within a narrow range around neutrality and remain very nearly constant for any given wastewater of a community. However, in the case of industrial plants and combined systems with both industrial and municipal discharges, it is not uncommon to encounter pH shock loadings.

Aerobic biological treatment processes can be operated efficiently within a reasonable wide range of pH values whereas anaerobic digestion systems, because of the sensitive nature of the methane forming bacteria operating in these systems, are significantly influenced by relatively minor pH fluctuations.

A. Aerobic Systems

Attached growth systems (e.g., biological filters), whether aerobic or anaerobic, are generally more resistant to pH variations than suspended growth systems, because of the following reasons: (1) higher levels of biomass concentration are expected to be present, (2) wash-out of microorganisms is usually less than in suspended growth systems due to their attachment to surfaces, and (3) there is a greater scope for microenvironments in thicker biofilms, where the pH could be favorable compared to the pH of the outside medium.

Among the suspended growth processes, cell recycle systems (e.g., activated sludge) are less sensitive to pH changes. Cell recycle provides higher biomass concentration and better opportunity for acclimatization of microorganisms to changing environmental conditions. It has been experimentally demonstrated that in cell recycle systems, a pH change from 6.7 to 3.2 had almost no effect on the concentration of organics in the effluent, while similar pH shock in once-through systems resulted in a significant loss of substrate in the effluent.[24] The shock, however, caused a drastic change in species predominance in both systems with a concomitant decrease in the settleability of biomass.

In continuous growth mixed culture once-through aerobic systems (e.g, aerated lagoons), the following observations were made, as a result of pH changes:[24] (1) increased concentration of substrate in the effluent; (2) wash-out of a large fraction of the biomass; (3) change in the species predominance; and (4) decreased settleability of cells due to enhanced filamentous growth. The prime mechanism of response to pH shock appears to be ecological, i.e., a shift in predominating species rather than biochemical acclimation of a large portion of the species present prior to the shock.[24]

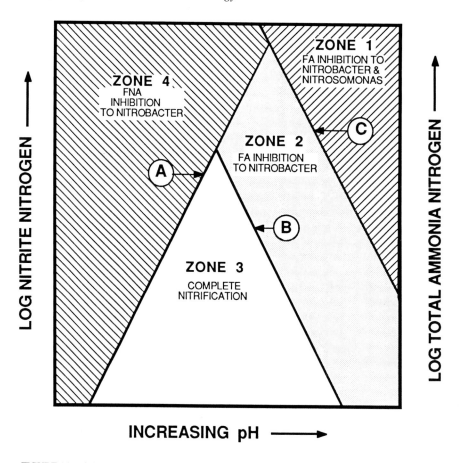

FIGURE 14.   Schematic representation of FA and free nitrous acid inhibition of nitrifying organisms.[10]

The effect of pH on the nitrification of sewage in aerobic processes has been reported to be variable and the optimum range appears to be between 7.0 and 8.5.[25] In a study in which the effect of pH on the oxidation of NH_4^+ and NO_2^- was studied with pure cultures and nitrifying oxidation ditch mixed liquors, a pH range of 7.4 to 7.9 was found to be optimum for both.[26]

Hydrogen ions are released during the oxidation of NH_4^+ and hence the pH drops in a nitrifying system.[27,28] The magnitude of the decrease in pH depends on the quantity of NH_4^+ oxidized and the buffer capacity of the wastewater concerned.[29] Inhibition of nitrification in wastewaters below and above the optimum pH for nitrification has been attributed to increasing concentrations of undissociated HNO_2 and free NH_3 occurring at progressively lower and higher pH values than 7, respectively.

A schematic representation of the relationships of FA and free nitrous acid inhibition of nitrifying organisms is presented in Figure 14. Zone 1 in the figure represents the condition when the FA concentration is high enough to inhibit both nitrosomonads and nitrobacters. No nitrification will occur and ammonia will accumulate in the system. At lower concentrations of FA, only nitrobacters may be inhibited and nitrite accumulation will occur. This condition is represented by zone 2. At still lower FA concentrations, neither nitrosomonads nor nitrobacters will be inhibited and complete nitrification will occur (zone 3). In the absence of any free nitrous acid inhibition, complete nitrification should represent conditions to the left of line B. Inhibition of nitrification by free nitrous acid may occur, and zone 4 represents the latter condition. The boundaries of the zones are noted as A, B, and C in Figure 14 as sharp separations.

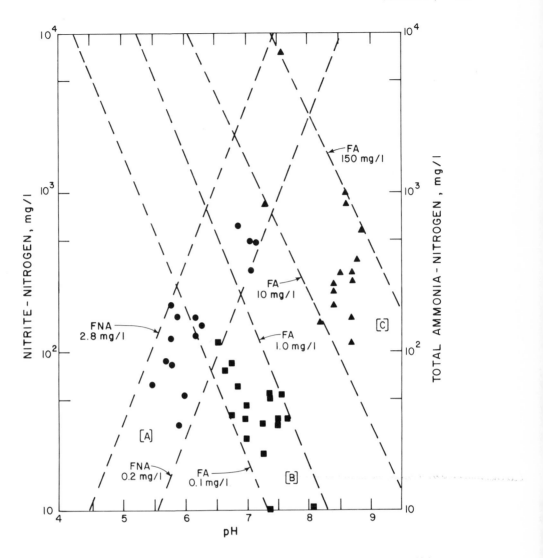

FIGURE 15. Experimental results showing FA and free nitrous acid inhibition of nitrifying organisms.[10]

A summary of the data collected from various experiments on nitrification is presented in Figure 15, which illustrates the boundary concentrations of different species of nitrogen at which they are inhibitory to nitrification. FA inhibition to ammonia and nitrite oxidizers appears to begin in the range of 10 to 150 mg/ℓ and 0.1 to 1.0 mg/ℓ, respectively. The inhibition of nitrite oxidizers appears to occur in the range of 0.22 and 2.8 mg of un-ionized $HNO_2\ell$ as depicted in Figure 15. The inhibition of nitrifiers by ammonia and nitrous acid has been found to be reversible in wastewaters by controlling the pH appropriately.[10] From an analysis of previously published experimental data, it has been reported recently that the optimum pH for ammonia-nitrogen oxidation varies according to its concentration in the medium (approximately 0.8 pH units between 1 and 50 mg/ℓ). However, the selected optimum pH value for the oxidation of ammonia nitrogen may not be optimal for the oxidation of the nitrite-nitrogen formed.[30]

B. Anaerobic Systems

The hydrogen ion concentration is one of the most important parameters in anaerobic

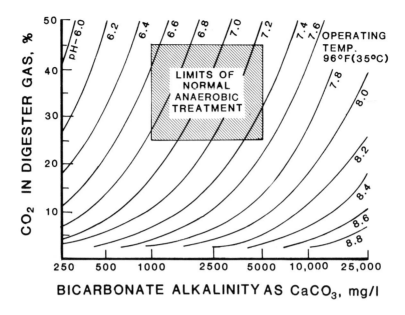

FIGURE 16.   Effect of pH and bicarbonate alkalinity of % CO_2 in digester gas.

systems. It influences the process performance and methane production in several ways: (1) volatile acids, which are usually the contributing factor for low pH conditions are toxic to methanogenic bacteria at high concentrations; (2) free ammonia, which is the predominant species over NH_4^+ at pH levels greater than about 9, is inhibitory to methane producing organisms; (3) pH and bicarbonate alkalinity, which are interrelated to each other affect the gas composition; (4) removal of heavy metals as precipitates (mostly as sulfides) is primarily a function of pH; and (5) availability of nutrients is influenced by the hydrogen ion concentration due to possible precipitation reactions rendering them unavailable for cell uptake.

In anaerobic systems such as sewage sludge digesters the relationship among volatile acids, alkalinity, and pH is of great importance, because bicarbonate alkalinity acts as a buffer against changes in pH due to volatile acid production.

The concentration of HCO_3^- can be expressed in terms of the percent CO_2 in digester gas phase by the following equation:

$$\frac{6.3 \times 10^{-4} p_{CO_2}}{10^{-pH}}$$

where p_{CO_2} is the partial pressure of CO_2. The interrelationship among the pH, bicarbonate alkalinity, and the CO_2 in the gas is presented in Figure 16. It can be seen from the figure that within a pH range of 6.5 to 7.5, the bicarbonate alkalinity should be maintained between 1000 to 5000 mg/ℓ in order to limit the CO_2 in the digester gas to a maximum of about 40%.

C. Chlorination

Although chlorination is a chemical process, the mechanism through which the treatment objective, namely disinfection is achieved is biological, and so considered here for discussion. pH is a key variable affecting the efficiency of chlorination process due to its influence on the relative concentrations of different chlorine species.

Generally, HOCl is considered to be a far more effective disinfectant than OCl^-. Several investigators[32,33] have demonstrated that the former is 70 to 80 times more bactericidal than

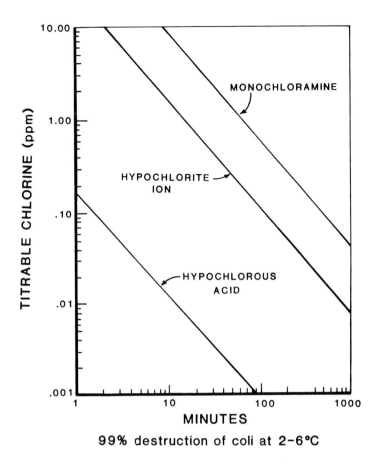

FIGURE 17.   Relative germicidal efficiency of HOCl, OCl⁻, and NH₂Cl[8]

hypochlorite ion. Thus, increasing the pH reduces germicidal efficiency, since OCl⁻ is the dominating species at high pH levels. Free chlorine residuals (OCl⁻ and HOCl) are more germicidal than combined chlorine residuals, i.e., chloramines.[8] The relative germicidal efficiencies of HOCl, OCl −, and monochloramine are presented in Figure 17. It has been experimentally demonstrated that dichloramine was a more potent germicide than monochloramine. Similarly, dichloramine was found to be more effective as a viricide[330] and a cysticide[34] than monochloramine.

VII. SUMMARY

pH is an important factor in wastewater treatment processes because of its influence on the process performance and efficiency. Among the several chemical equilibria that exist in the wastewater process streams, carbonate, ammonia, sulfide, and chlorine systems are considered the most important, and the relative distribution of different species in each one of these systems is primarily a function of the pH. Biological processes may be influenced by hydrogen ion concentration in several ways: (1) different groups of microorganisms may become dominant at different pH levels; (2) extremely low or high hydrogen ion concentrations may be directly inhibitory to the enzymatic activity of the microorganisms; and (3) inhibition (or enhancement) of biological activity may be caused by certain chemical species, whose ionic state is altered as a result of change in the pH of the system.

Most bacteria grow best under neutral or slightly alkaline conditions, but yeasts and fungi prefer an acid environment. Many algae are tolerant of mild acidity, whereas most blue green bacteria have pH optima on the alkaline side of neutrality. Among the common biological treatment processes, nitrification and anaerobic systems are the most sensitive to pH fluctuations.

REFERENCES

1. U.S. Environmental Protection Agency, Phosphorous Removal. Process Design Manual, USEPA, Washington, D.C., 1971.
2. **Patterson, J. W., Allen, H. E., and Scala, J. J.,** Carbonate precipitation for heavy metal pollutants, *J. Water Pollut. Control Fed.,* 49, 2397, 1979.
3. **Nemerow, N. L.,** *Liquid Wastes of Industries: Theories, Practices, and Treatment,* Addison-Wesley, Reading, Mass., 1971.
4. U.S. Environmental Protection Agency, *Sludge Treatment and Disposal. Process Design Manual,* EPA 625/1-79-011, USEPA, Washington, D.C., 1979.
5. **Gossett, J. M., Healy, J. B., Jr., Stuckey, D. C., Young, L. Y., and McCarty, P. L.,** Heat Treatment of Refuse for Increasing Anaerobic Biodegradability, Dept. of Civil Engineering Tech. Rep. No. 205, Stanford University, Stanford, Calif., 1975.
6. **Elliott, H. A. and Huang, C. P.,** Adsorption characteristics of some Cu (II) complexes on alumino silicate, *Water Res.,* 15, 849, 1981.
7. **O'Melia, C. R.,** Coagulation, in *Water Treatment Plant Design for the Practicing Engineer,* Sanks, R. L., Ed., Ann Arbor Scientific, Ann Arbor, Mich., 1978, 65.
8. **White, G. C.,** *Handbook of Chlorination,* Van Nostrand Reinhold, New York, 1972.
9. **Alexander, M.,** *Introduction to Soil Microbiology,* 2nd ed., John Wiley & Sons, New York, 1977.
10. **Anthonisen, A., Loehr, R. C., Prakasam, T. B. S., and Srinath, E. G.,** Inhibition of nitrification by ammonia and nitrous acid, *J. Water Pollut. Control Fed.,* 48, 835, 1976.
11. **McCarty, P. L.,** Anaerobic waste treatment fundamentals. III. Toxic materials and their control, *Public Works,* 11, 91, 1964.
12. U.S. Environmental Protection Agency, Sulfide Control in Sanitary Sewerage Systems. Process Design Manual, USEPA, Washington, D.C., 1974.
13. **Sano, H.,** The role of pH in the acute toxicity of sulfite in water, *Water Res.,* 10, 139, 1976.
14. **Leeper, G. W.,** *Managing the Heavy Metals on the Land,* Marcel Dekker, New York, 1978.
15. **Snoeyink, V. L. and Jenkins, D.,** *Water Chemistry,* John Wiley & Sons, New York, 1980.
16. **Stumm, W. and Morgan, J. J.,** *Aquatic Chemistry,* 2nd ed., Wiley-Interscience, New York, 1981.
17. **Pipes, W. O.,** pH variation and BOD removal in stabilization ponds, *J. Water Pollut. Control Fed.,* 34, 1140, 1962.
18. **Desrosier, N. W.,** *The Technology of Food Preservation,* AVI, Westport, Conn., 1970.
19. **Pelczar, M. J., Jr. and Reid, R. D.,** *Microbiology,* McGraw-Hill, New York, 1958.
20. **Gaudy, A. F., Jr. and Gaudy, E. T.,** *Microbiology for Environmental Scientists and Engineers,* McGraw-Hill, New York, 1980.
21. **Hamilton, W. A.,** Energy coupling in microbial transport, *Adv. Microb. Physiol.,* 12, 1, 1975.
22. **Brock, T. D.,** *Biology of Microorganisms,* Prentice-Hall, Englewood Cliffs, N.J., 1970.
23. **Atlas, R. M. and Bartha, R.,** *Microbial Ecology: Fundamentals and Applications.* Addison-Wesley, Reading, Mass., 1981.
24. **George, T. K. and Gaudy, A. F., Jr.,** Response of completely mixed systems to pH shock, *Biotechnol. Bioeng.,* 15, 933, 1973.
25. **Wild, H. E., Sawyer, C. N., and McMahon, T. C.,** Factors affecting nitrification kinetics, *J. Water Pollut. Control Fed.,* 43, 1845, 1971.
26. **Srinath, E. G., Loehr, R. C., and Prakasam, T. B. S.,** Nitrifying organism concentration and activity, *J. Environ. Eng. Div. ASCE,* 102(EE2), 449, 1976.
27. **Loehr, R. C., Prakasam, T. B. S., Srinath, E. G., and Joo, Y. D.,** Development and Demonstration of Nutrient Removal from Animal Wastes, Environ. & Technol. Ser. EPA-R2-T3-005, Office of Research and Monitoring, USEPA, Washington, D.C., 1973.
28. **Painter, H. A.,** A review of literature on inorganic nitrogen metabolism in microorganisms, *Water Res.,* 4, 395, 1970.
29. U.S. Environmental Protection Agency, Nitrogen Control. Process Design Manual, USEPA, Washington, D.C., 1975.

30. **Prakasam, T. B. S. and Loehr, R. C.,** Microbial nitrification and dentrification in concentrated wastes, *Water Res.,* 6, 859, 1972.
31. **McCarty, P. L.,** Anaerobic waste treatment fundamentals. II. Environmental requirements and control, *Public Works,* 10, 123, 1964.
32. **Fair, G. M., Morris, J. C., Chang, S. L., Weil, I., and Burden, R. P.,** The behavior of chlorine as a water disinfectant, *J. Am. Water Works Assoc.,* 40, 1051, 1948.
33. **Fair, G. M., Morris, J. C., and Chang, S. L.,** The dynamics of water chlorination, *J. N. Engl. Water Works Assoc.,* 61, 285, 1947.
34. **Kelly, S. M. and Sanderson, W. W.,** The effect of chlorine in water on enteric viruses. II. The effect of combined chlorine on poliomyelitis and coxsackie viruses, *Am. J. Publ. Health,* 59, 14, 1960.

INDEX